Out of Its Mind

OUT of ITS MIND

Psychiatry in Crisis

A CALL FOR REFORM

J. Allan Hobson
Jonathan A. Leonard

PERSEUS PUBLISHING
Cambridge, Massachusetts

Copyright © 2001 by J. Allan Hobson and Jonathan A. Leonard

Cataloging-in-Publication Data is available from the Library of Congress

ISBN 0-7382-0251-7

Perseus Publishing is a member of the Perseus Books Group.

Find us on the World Wide Web at http://www.perseuspublishing.com

Perseus Publishing books are available at special discounts for bulk purchases in the U.S. by corporations, institutions, and other organizations. For more information, please contact the Special Markets Department at HarperCollins Publishers, 10 East 53rd Street, New York, NY 10022, or call 1-212-207-7528.

Text design by Heather Hutchison
Set in 11.5-point Bulmer by Perseus Publishing Services

Illustrated by Ellen Edwards

First printing, April 2001
1 2 3 4 5 6 7 8 9 10—03 02 01

To Our Fathers:
John Robert Hobson and Jonathan N. Leonard

Contents

List of Figures

Acknowledgments

We have done everything possible to avoid building this book like the Tower of Pisa—on a narrow base of information in uncertain terrain. Instead we have tried to build it like the Eiffel Tower—believing it would need height to effectively scan an extensive landscape, and seeing the need for a broad base of knowledge firmly anchored in many disciplines.

To help prepare the way, we began by interviewing a wide range of experts—people who are familiar with psychiatry's past and present, concerned about current issues, and in good command of some relevant field of knowledge. Four of those interviewed work at the National Institute of Mental Health (NIMH) in Washington, D.C. They are Steven Hyman, the director, whose specialty is genetics; Daniel Weinberger, a noted leader in schizophrenia research; Philip Gold, an expert on depression; and Judith Rapoport, a distinguished child psychiatrist.

Many of the others interviewed are affiliated with Boston's Massachusetts Mental Health Center. They include Ming Tsuang, the center's director and an authority on the genetics of schizophrenia; Carl Salzman, a leader in psychopharmacology; Alan Green, an expert in atypical antipsychotic medication; Joseph Schildkraut, a pioneer in research on depression; Lester Grinspoon, another research pioneer noted for work on schizophrenia; and Thomas Gutheil, an expert on psychiatry and the law. In addition, we interviewed two people at the center who do research in Allan Hobson's Laboratory of Neurophysiology. One of them, Robert Stickgold, helped to assess psychiatry's current problems relative to advances in brain science research; and the other, Edward Pace-Schott, provided insight into how psychiatry's problems look from the perspective of therapeutic counseling.

We also interviewed two other people, one in England and the other in Washington, D.C. The first was Mark Solmes, a psychiatrist at the Royal Hospital of London who is seeking to bridge the gap between psychoanalysis and brain science. And the other was E. Fuller Torrey—schizophrenia researcher, administrator, authority on psychiatry's history, and a leading social critic of psychiatry's current plight.

Not all the people interviewed are quoted in the text; but all of them enriched the book's content, and we are grateful to all for their assistance.

Regarding primary scientific information contributing to the book's theoretical structure, Allan Hobson's research has been generously supported by grants from the National Institutes of Mental Health (NIMH 13,923 and NIMH 48,832), the National Institute of Drug Abuse (NIDA, NIH-IR01DA11744 01A1), and the National Aeronautics and Space Administration (NASA, NAS 9-19406). Such scientific progress as has been made is the fruit of collaboration with many colleagues, students, and patients who have endeavored to advance our understanding of the brain and its mind over the past forty years. The Laboratory of Neurophysiology, where much of this research has taken place, is located in the Massachusetts Mental Health Center, which has benefited greatly from its support by the Commonwealth of Massachusetts and its affiliation with the Harvard Medical School. Dr. Hobson's interdisciplinary impulses have been directly encouraged by his decade of interaction with fellow members of the Mind–Body Network of the John D. and Catherine T. MacArthur Foundation.

As coauthors we are grateful to *Harvard Magazine* for bringing us together in the first place and to Amanda Cook of Perseus Books for her able editorial guidance. We are also grateful to Robert Goisman, Lester Grinspoon, Carl Salzman, Joseph Schildkraut, and Ming Tsuang for critiquing portions of the text; to Constance Burr at NIMH for helping to arrange interviews; to Bruce Rheinstein at the National Alliance for the Mentally Ill for assistance with statistical data; to Dawn Opstad and Michael deNero at the Laboratory of Neurophysiology for their coordinating efforts; and to our wives Rosalia and Jeanne for their support and encouragement in helping this book come to light.

PART ONE

Psychiatry's Lost Mind

1

The Unfinished Revolution

*The era of Freud and the era of deinstitutionalization are
ending. The era of neurobiology has begun and promises to
revolutionize our understanding of the brain and its diseases.
For those of us who are interested in mental illness, it is an
exciting, perplexing, chaotic time, a time of crisis that
demands confrontation.*

—**E. Fuller Torrey,**
***Out of the Shadows,* 1997**

Clearly, psychiatry is in crisis. Nobody denies that. Psychiatrists bemoan
their field's fragmented, underfinanced services. Patients and coworkers
worry about numerous foreign psychiatric residents who can barely speak
English. And we can all see hordes of severely disordered people being
consigned to homeless shelters and to prisons, while private hospitals commonly turn away such people and mental hospitals continue to shut down.

So the question is not whether psychiatry confronts a crisis but what can
be done about it. Our purpose here is to offer a sound answer. That answer
isn't obvious, because an obvious answer would already be at hand. But it's
clear enough if the background is filled in. So our first task is not to make
proposals but to fill in that background—by explaining how psychiatry
came to lose its way.

A good place to start is with the coming of strong and effective psychiatric medication. That's because, from the days of Sigmund Freud to the present, few events in the history of psychiatry have sparked such drastic change as the emergence of psychiatric drugs. These drugs first made medical news in the 1950s, when millions of the mentally ill were still being warehoused in asylums and when Freudian psychoanalysis was coming to hold undisputed sway at most U.S. universities, mental hospitals, private offices, and clinics. Slowly, over several decades and in combination with certain questionable social theories about how easy mental illness was to treat, the new drugs changed all that. No matter that they did not cure but merely moderated target ills, or that they had disagreeable and dangerous side effects, or that their benefits were sometimes overstated, or that only some patients responded well to them. The plain fact is, they offered hope as nothing else in psychiatry ever had. They offered a real chance for large numbers of severely afflicted mental patients to have at least a semblance of normal life. So they were embraced with a mounting medical fervor that came to shake the field of psychiatry like an earthquake, trapping the Freudians in their own theoretical constructs, splitting open and dumping out the mental hospitals, and fracturing the entire field along fault lines still visible today.

Unfortunately, the revolution thus unleashed was never finished. The Freudians became isolated and dethroned but unrepentant, while psychiatry wavered about what besides drugs should be stuck into the yawning therapy gap the dethroned Freudians left behind. But the insurance companies and HMOs that gained authority in the 1980s and 1990s didn't waver. They simply downsized psychiatry, replacing high-paid psychiatrists with low-paid therapists and counselors, and turning those psychiatrists they continued to employ into pill-pushers who hardly spoke with patients—to a point where the best medical school graduates no longer wanted to become psychiatrists.

Blown hither and yon by these trends, the people with severe mental ills who had been psychiatry's chief wards fared poorly. Most were unceremoniously decanted out of America's mental hospitals, mainly in the 1960s and 1970s, and the doors of most mental hospitals were closed behind them—with assurances that better care in the form of community support services would be provided. With few exceptions, however, these community services failed to materialize, leaving the bewildered multitudes to their own devices. Small wonder, then, that many of the people abandoned in this fashion wound up in nursing homes or jails; or that we now have over two million Americans with severe but untreated mental ills; or that the plight of the severely disordered homeless on our streets has become a socially indefensible disgrace.

More broadly, far greater multitudes with lesser ills are also treated poorly. That's because most therapists cannot prescribe drugs; most drug-prescribing psychiatrists have only brief contact with their patients; and communication between therapists and physicians, who are often engaged in turf wars, is poor or absent. So all too often the process of diagnosis, therapy, and drug prescription becomes fragmented, and patients seeking well-coordinated care find it does not exist.

Meanwhile, ironically, the drugs that provided the impetus for all this have improved. They have become more versatile, less dangerous, and better able to treat a wider range of mental ills, including schizophrenia, manic depressive illness, depression, and many anxiety disorders. Beyond that, as shall be seen, we have made enormous strides in understanding how the brain works, how our drugs operate, what causes various kinds of mental illness, and what can be done to improve both drugs and treatment.

All of this suggests that the time has come to consolidate our gains and complete psychiatry's unfinished revolution. But old ways die hard. In World War I we had plenty of battlefield commanders yearning for horses instead of tanks, and we face a similar situation in psychiatry today. So we need to update and revive psychiatry.

One way this should be done is with a new psychology—a psychology that will harness brain science knowledge to the task of advancing our understanding of the mind. As will be seen later, we have tried to frame such a psychology. In so doing, we have found that the "psychodynamic" psychology commonly used by psychiatrists today is useful in one way but deficient in another. It is useful because it stresses the dynamic nature of mental processes; but it is deficient in that it treats the mind or "psyche" like an independent entity not subject to scientific assessment.

So how might we replace the term "psycho" in pyschodynamics? We could of course replace it with the word "brain." After all, we are using the term "brain science" throughout this book, rather than the more traditional "neuroscience," to refer to the study of neural processes, because scientific study of neural processes has long since passed the point where it is relegated to probing individual neurons and is now earnestly engaged in studying that vast mass of neurons we call the brain. Even so, the term "brain dynamics" seems awkward; and since the "neuro" prefix is still used freely in other terms like "neurology" and "cognitive neuroscience," we have chosen to call our new psychology "neurodynamics."

Those aware of our devotion to brain science may suspect that neurodynamics will prove to be something of a Trojan horse—that we are really pressing for "more biomedicine" or "more drugs." They could hardly be more wrong. The truth is that psychiatry has developed something of a split personality, with some practitioners pushing pills at the behest of HMOs while others cling to outmoded forms of psychology divorced from medicine. As a result, psychiatry needs to renew itself by healing this rift, harmonizing its psychology with our growing knowledge of brain science, developing a comprehensive treatment system that will keep people with severe mental ills from being neglected, and overseeing sound treatment for others with lesser ills. This, as opposed to overzealous pursuit of impersonal biomedicine, is the right path to sound treatment.

Anyone who doubts this fact need only recall that virtually all our psychoactive drugs are limited. They just treat symptoms. They don't cure the underlying disease. So even if the patient responds ideally, the best a drug can do is to banish or reduce the patient's symptoms. And if the patient stops taking the prescribed medicine—because of side effects, poor judgment, some personal crisis, reduced monitoring, or any other reason—the symptoms may return. Also, people vary a good deal and so do mental ills, creating a clear need to make decisions on an individual basis. As a result, the act of prescribing psychiatric drugs is not so much a science as an art: an art that needs to be interwoven with monitoring, therapy, the ups and downs of the ailment being treated, changes in the patient's situation, and various kinds of community and family support.

Sadly, in the current era of managed care we commonly see only a parody of such enlightened treatment. For the HMOs and insurance companies have tended to assign therapy, the psychological side of patient care, to well-meaning but biologically unsophisticated psychologists and counselors. Intensifying the irony, many of these caretakers are steeped in psychodynamics or some other type of therapy divorced from medicine. So at some critical point the poor patient is apt to be shuttled back and forth between the two poles of psychiatry's split personality—from the pill-pusher's "mindless" pharmaceutical methods to the caretaker's "brainless" psychodynamic methods. Need we wonder why so many patients are lost to treatment?

What deserves particular attention is not so much the shuttling around but the fact that vast numbers of patients fail to get the comprehensive care they need. As an example of this general problem, consider the case of someone we shall call Alice Morrisey, a businesswoman who consulted one of us (Allan Hobson) for help with a sleep problem. Her name, like those of all mental patients mentioned in this book, is fictitious. But her plight is not fictitious, and her story poignantly underlines the need for a coordinated program combining support for psychotherapy with versatile pharmacology, temporary hospitalization, and social services. Unfortunately, even for such high-functioning patients, programs of this sort are hard to find.

Alice Morrisey's sleeping difficulties began in January 1997. Prior to seeing me intermittently (a total of nine times from December 1997 through May 1998), she had acted as the caretaker for an elderly man who was in poor health and who subsequently died. During her period of employment (October 1996–February 1997) she came into conflict with the man's son, whom she accused of elder abuse and of threatening her sexually and physically. She said that during one disagreement with him he came into the room in his underwear and showed her pictures of nude young women fondling each other. She claimed this incident had triggered both anxiety and insomnia in her because it challenged her integrity as a concerned caretaker and aroused fears of physical and sexual assault. Alice was so convinced that her sleep disorder was triggered by this encounter that she had brought suit against her employer.

She reported waking up four to seven times per night, often in a sweat, following dreams that involved intense anxiety. The dreams' content varied but included the sensation of being nibbled by animals, a feeling of being pursued, fear of dying, and terror that the world was ending. People in the dreams included her mother, the old man she was taking care of, and the man's son. She said she often awoke to find her pillow wet with tears. By her account, the resulting lack of sleep caused lapses in her concentration, frequent daytime naps, and chronic fatigue. She also reported having anxiety attacks in the daytime, with palpitations and sweating.

Alice's sleep disorder did not respond to supportive psychotherapy or to a wide array of drugs, including Ambien, Ativan, Buspar, Depakote, Prozac, Stelazine, and Trazodone. However, when she was seeing me on a semiregular basis she seemed to be holding her own despite the sleep and anxiety disorders.

In order to understand her current crisis and to respond sympathetically, I became familiar with Alice's past history. She was born in 1941 and was the youngest of six children. Her mother was a stalwart Irish Catholic and her father was a painting contractor. She reported that she admired her father for his music and poetry but often had to rescue him from barroom blackouts and the DTs. A bright girl, she exhibited no obvious mental ills in her early years but was prone to anxiety. This flared up when she was 18, after she was involved in a car accident at college. On that occasion she was prescribed a sedative (glutethimide), to which she became temporarily addicted. Both of her parents died in the 1970s.

Anxiety and conflict often surfaced in Alice Morrisey's relationships with men. Her first marriage lasted seven years, her second five. A third long-term relationship that endured for eight years ended in 1984. During her first marriage she had a daughter, Vanessa, who became a state beauty queen in 1988.

Beginning around 1990, when she was 49 and menopausal, things took a downward turn. Her daughter married a man of whom she did not approve; her previously successful travel business started to fail; a three-year relationship with a man broke up; and two of her sisters and a brother became ill and died of cancer. In November of 1992 she was treated for anxiety at a teaching hospital near her home. At that time she reported the belief (a delusion) that her sisters were not dead.

A short time later, in the winter of 1992–1993, she began acting grandiose, stopped eating, stopped taking medication, and became frankly psychotic. This was an understandable but regrettable development that might have been prevented had she been under the continuing care of a psychiatrist. Her behavior then got so bizarre that the local police arrested her. They found her to be confused and delusional and took her to a general hospital. According to her own account she was diagnosed there as having autism and schizophrenia, was placed on the drug Haldol (an antipsychotic used mostly against schizophrenia), and was discharged. Later, at a state psychiatric hospital, the diagnosis was changed to major depression, she was switched to Prozac (used mostly against anxiety and depression), and her condition gradually improved.

After that she worked at a series of caretaker jobs and struggled unsuccessfully to return to her once-successful travel business. Around June 1996, following a move to another town, her psychosis resurfaced. This time her grandiose behavior and delusions included the paranoid belief that her friends were working for the FBI. She was committed to another state mental hospital for a month and then discharged, being seen intermittently by a psychiatrist for about a year thereafter. From the time of her discharge until her most recent hospitalization she functioned erratically on a daily dose of 200 mg of Tegretol (carbamazepine), a drug sometimes used for mood stabilization.

A major problem for Alice was where to turn when she was discharged from hospitals. Not having any nearby siblings who could help her, she had developed several patron caretakers. There was a group of nuns she sometimes lived with; and a couple she had provided care for who were devoted to her; and an old alcoholic college friend married to a motel owner who would let her stay in the motel and work in the front office. Not having any money for an apartment, she would bounce around from one of these marginal living situations to another. It was in this context that she took the elder care job that precipitated her current crisis and resulted in her deciding to take legal action against her employer and to consult me.

After May 1998, when her visits to me had ended, every so often she'd have anxiety-related troubles and would phone me. I would talk to her and try to piece things together as best I could. Obviously, this was a marginal arrangement. In July of 1998 she became very disorganized and was hospitalized briefly. Then, over a year later, in the fall of 1999, she lost contact with me altogether. In December 1999, after her lawsuit had been settled out of court, she became floridly psychotic—so psychotic that there was simply no talking to her. She was admitted to a general hospital just before Christmas, and was later transferred to a state mental hospital.

Alice Morrisey has clearly been suffering from a major affective (emotional) disorder with intermittent manic-psychotic episodes. However, her case has a number of striking features. To begin with, anyone who met her in a lucid period (I never saw her when she was psychotic) would be impressed. This lady ran a flourishing business. She was smart and industrious. Also, while she had long experienced anxiety-related difficulties, probably at least since her college days, her florid symptoms emerged very late, so late that she was never psychotic before the age of 50.

Another unusual feature was her poor response to drugs. Aside from her apparently favorable experience with Prozac in 1993, most of her other responses to psychiatric medications ranged from poor to nil. I'm not clear why she was never given lithium, the principal medication used against mania, but there are a lot of drugs she either couldn't or wouldn't take. That probably goes a long way toward explaining why most of the drugs she did take, coming as they did from a restricted list of choices, seem to have done her little or no good.

Another big problem was that she had no family nearby to look after her. Because today's psychiatric care is so badly organized, parents or siblings typically pick up the pieces, seeing that the care given more or less matches what's needed, and making sure the patient doesn't reach a point where he or she becomes suicidal, assaultive, homeless, or imprisoned—the usual fate of today's mentally ill people who hit bottom.

But Alice Morrisey had no desire to fall through the cracks, and she had jury-rigged a social support network of her own that operated after a fashion. She also had good "insight" into her illness. That is, even when psychotic she appears to have understood that she needed help. So she always let herself be hospitalized, which at least kept her from being turned out to wander in the streets. Nevertheless, while she was outwardly compliant and even ingratiating, Alice's judgment was impaired even when she was free of psychosis, pointing to the possible existence of an underlying personality disorder.

Though Alice Morrisey's situation may improve, her recent history stands as a stern indictment of today's psychiatric care. In 1992–1993 she took three trips through hospital revolving doors before being put on a drug that seemed to help. And from 1993 onward it seems clear that she needed sound and coordinated diagnosis, pharmacology, therapy, and follow-up. Time and again she got one or another of these things but not all of them together.

If she had managed to establish a sound, long-term relationship with a physician she trusted—one who knew how to use both drugs and psychological techniques, and who was part of a cohesive system able to follow up her case—she might have avoided much of her later trouble. That doesn't mean that either drugs or psychotherapy could have made her mental illness go away. That's too strong a claim for someone with what are almost certainly lifelong problems. But proper care could have helped her to stay functional.

Even quite late, when she was seeing me on a semiregular basis, she was doing pretty well. Of course, I was merely being consulted for her sleep disorder. When her money ran short, she couldn't fix her car and couldn't come to Boston. And since then, obviously, she's gone from bad to worse. It's speculative to assume that we could have countered these processes just through weekly or biweekly office visits, though I think that's fairly likely. But in any case, that didn't happen. Instead, she didn't get what she thought she needed, she didn't get what I thought she needed, and she slid down the back stairs. Her distressing case shows why it is important to get long-term help from someone who understands both drugs and therapy, and also points up the urgent need for a strong support structure backing up the psychiatrist—one that will provide patients with medical monitoring, psychotherapy, case management, community support, and follow-up.

A prime reason why effectively coordinated services like these are not the norm is simply that people with severe mental disorders lack clout. Poverty, incoherence, disorganization, and social stigma all tend to pull them down. And since they cannot do much to stand up for themselves, lawyers, sociologists, and others with their own agendas have in the past been free to concoct high-sounding social arguments that work against them. For similar reasons, politicians, administrators, and other social leaders tend to shun them—or what is worse, tend to uncritically accept ideas like the infamous "patients' rights" argument against commitment offered up by poorly informed lawyers and social scientists. With such things going on, it is easy to see why vast multitudes of the severely ill have been neglected.

Of course, anyone who seeks to change a situation should have a good grasp of what that situation is. So we will take time in the next chapter to describe today's mental health scene, focusing mainly on severe disorders and on people whose treatment has ranged from inadequate to nil. We will also focus on psychiatry's role in all this and will try to explain why, in the postasylum era, it has failed to protect its former charges.

Part of the answer can be traced to psychiatry's old affair with Freudian psychoanalysis—a promising relationship that ultimately proved ill-suited to treating severe ailments. But it is unfair to simply blame the Freudians for what went wrong. So it is worth telling psychiatry's story: how it was originally tied to the state mental hospitals; how it emerged to embrace psychoanalysis; and how the future of psychoanalysis seemed so bright in the 1950s and 1960s that the best medical school graduates were attracted to psychiatry like bees to honey.

We shall also see how the emergence of effective drugs joined up with several things—public distaste for the mental hospitals, waning confidence in psychoanalysis, and confusion about the nature of mental ills—to precipitate psychiatry's decline. We shall watch as psychiatry's pendulum swings from the brainless mind of Freud to the mindless brain of biomedicine. We shall assess the current crisis, how psychiatry operates, and rising public discontent with the way so many severely ill people are being treated. And we shall suggest that the prospects for reforming and reviving psychiatry are brighter than one might think—because the public wants sound treatment of mental ills; such treatment is now feasible if the gap between psychotherapy and biomedicine is bridged; and bridging that gap is precisely what a reformed psychiatry could do.

Something that makes these prospects especially good is our growing knowledge of how psychiatric drugs actually work. For the antipsychotic drugs that began emerging in the 1950s (the so-called "neuroleptics") were quite specific. They did not simply "dope up" the recipient until he or she became compliant. Rather, they targeted particular diseases, or even particular groups of disease symptoms.

Such specific effects were mysterious half a century ago, because the early neuroleptics were discovered by hit-or-miss research, and how they worked was uncertain. While not all this uncertainty has been banished, as time passed investigators came to realize that the drugs' specific effects reflected specific actions in the microscopic gaps (synapses) between nerve cells.

Most nerve cells (neurons) depend on chemical transmitters to send messages across those synapses. A transmitter chemical, released into the gap by the sending neuron, crosses the synapse and "docks" on the other side at receptor sites designed to accept it, thereby providing communication across the gap. But some chemicals called neuromodulators do far more than this. Instead of transmitting momentary signals, they alter the receiving neuron's receptivity to further signals, thereby modulating that neuron's pattern of activity. What the neuroleptic drugs were doing was blocking some of the "docks" available to receive a particular neuromodulator called dopamine. The net effect was to tone down certain brain activities, including activities causing various psychotic symptoms.

This model is worth noting, because most psychiatric drugs act similarly. That is, most of them promote or hinder passage of certain neuromodulators between neurons. This can be done various ways. It can be done by stepping up or reducing the amount of neuromodulator released into the synapse, or by changing the rate at which that chemical is broken down or removed from the synapse, or by blocking or otherwise changing the receptivity of docking sites on the receptor neuron.

This last method—altering or blocking docking sites—seems especially promising for drug development right now because there are many different kinds of docking sites. Indeed, there seem to be five or more different sorts of docking sites per neuromodulator for most of the brain's many neuromodulators, with each type of docking site tending to be distributed differently within the brain. A particular psychiatric drug, such as Thorazine (chlorpromazine), tends to act at one or more specific docking sites while ignoring other sites, and that makes its effects differ from those of other drugs. What's even more encouraging is that many of the newer drugs alter multiple docking sites for several neuromodulators in varying degrees. So in theory, if we can learn enough about the brain, there is good reason to suspect we might develop an array of versatile drugs to offset mental disorders by influencing one or more selected docking sites in particular regions of the brain, much as a piano tuner can offset disorders of that instrument by tuning certain well-selected strings.

There is room for optimism here, even though the human brain is billions of times more complex than a piano, because our current knowledge of the brain is growing fast. As the foregoing suggests, in recent decades we have learned volumes about neurons, neurotransmitters, and how nerve cells talk to one another. More generally, we have learned about brain chemistry and how such chemistry can direct the brain to sleep, dream, and awaken, as well as to experience bizarre effects like irrational panic or hallucinations. And we are learning about genes—how certain genes may raise intelligence, for example, or increase vulnerability to alcohol and drugs, or make us prone to schizophrenia, depression, and other mental ills.

Moreover, we now have various tools for penetrating the skull's bony X-ray barrier, so we can see living substructures within the human brain and watch the mind at work. Despite various limitations, these tools have powerfully complemented animal studies and other older methods, and so we have begun to pin down what key brain structures really do. We have begun to understand how the brain is coordinated, how emotions operate, how memories form, and even how consciousness may work. We have also developed compelling theories about the causes of many mental ills—including schizophrenia, depression, and various anxiety disorders; and while this burgeoning brain science cannot yet provide full answers to many questions about the mind, it has come far enough in a short time to convincingly demonstrate a fast pace of discovery and sunny prospects.

To some, this may seem more a curse than a blessing. Astrology and many New Age fashions demonstrate that people love mystery as well as knowledge, and not everyone wants to see the mysteries of the human mind unraveled. Also, there are sometimes religious reservations, and sometimes an antipathy toward science that shows up in things like the "creation science" policies that have worked their way into many of our schools. Collectively, these inclinations probably account for the attention garnered by people like John Horgan, whose books (*The End of Science* and *The Undiscovered Mind*) claim that neuroscience is pretty much in the dark about the mind and likely to remain so.

We disagree. As we see it, Horgan's examination of brain science is not just confused but misdirected. It is not a matter of the cup being half full or half empty, because in fact the cup is brimming over with discovery. But merely talking about this in general terms won't get the point across. Moreover, we like the idea of explaining where brain science is now—because that story is exciting, and because it shows how our ability to understand the mind and treat severe mental illness is improving. So in the middle portion of this book we will describe what is happening as best we can in plain language. We will explore the activities of neurotransmitters, brain cells, the brain's design, and how various brain structures interact. We will see what genetic and other research is telling us about brain development and how that development can go awry. We will examine the nature of consciousness, thought, sleep, dreams, and other exceptional mental states; probe the causes of shifts from dreaming to waking; tell more about brain chemistry and drugs; and set forth the basics of what biomedicine can do to treat severe disorders. For obvious reasons, our coverage of this vast medical and scientific territory will necessarily be brief. Even so, we expect to touch on many of the major advances in current knowledge and to carefully set apart things applicable now from insights and theory alive with promise for the future.

That still leaves us short of where we want to go. For psychiatry's neglected multitudes cannot be treated with raw knowledge. And as the availability of biomedical knowledge demonstrates, the main problems now are not biomedical but psychological and social. So we shall make proposals about how to inspire public confidence; build a new neurodynamic psychology; bridge the gap between biomedicine and personal therapy; devise a community-based structure to detect, treat, and monitor those who are being neglected; and inspire everybody from counselors to psychiatrists, medical students to professors, and legislators to general readers with true awareness of what can in fact be done. That is the aim of the last part of our book.

All this seems quite ambitious. But we are reminded of a time, barely a hundred years ago, when little could be done to prevent or cure scourges like malaria, yellow fever, cholera, typhoid fever, diphtheria, pneumonia, tuberculosis, and many others. In those years the germ theory of disease was advancing; Louis Pasteur and other scientists were learning about the microbial world, aseptic technique, immune sera, bad water, insect vectors, and how disease organisms get around. Indeed, a sea change in the rudimentary medical knowledge of the day was in progress. And while the new knowledge was imperfect (sterile techniques were poor, anesthesia uncertain, antibiotics unknown), when this knowledge finally got applied with vigor—most notably by U.S. occupation forces in Cuba, then in Panama, and then in World War I—it gave birth to what we know as "modern" medicine, banishing yellow fever from Cuba, permitting construction of Teddy Roosevelt's Panama Canal, saving hundreds of thousands if not millions of American lives in World War I, providing a sound foundation for future knowledge, enhancing general welfare in the United States and Europe, and advancing the average person's life expectancy many years.

Today's psychiatry is about where communicable disease medicine was then. Until recently we could do little to combat severe disorders of the mind. We had no idea what caused them; our asylums were little more than holding tanks; and Freudian progress was disappointing.

Of course, our hundred-year delay is understandable, because the human brain is hard to fathom and far tougher to study than most germs. Nevertheless, in recent decades we have been making progress. The past has given us extensive experience with patient monitoring and therapy; we have drugs that work reasonably well; we have an explosive advance in brain science knowledge; and we can see the beneficial results obtained when these things are applied together.

Therefore, it takes no special talent to see the possibilities—to see that we should press ahead vigorously, take command of our new knowledge, convincingly bridge the gap between biomedicine and therapy, set up community services, and reach the multitudes who are ill-served or neglected. To press for such change now may seem a bit precipitous; but all the preliminary knowledge needed is in place. What's more, we have a precedent. For in fact we are doing nothing more than taking a leaf out of medical history and urging social progress similar to that accomplished by William Gorgas, William Welch, and other health pioneers around the dawn of the twentieth century, when modern medicine was young.

We admit that talk is cheap. The steps taken by our early health pioneers weren't easy, and we don't expect the ones we propose here will be easy. But they are feasible. They are morally acceptable. They come at the right time because public patience is wearing thin. And they hold promise of ending a pattern of social neglect increasingly regarded as cruel, wasteful, and even dangerous. So there is good reason to suspect that our call to action will get the attention it deserves and will be seen correctly as something that goes beyond ourselves, being no mere formulation of clever thoughts by the authors, but instead being one early embodiment of an idea whose time has come.

2

Out of Bedlam

If the American public understood the reality of the hundreds of thousands of [mentally ill] children and adults who are caught in the criminal justice system in this country because they don't have access to community-based services, and saw the kind of suffering and degradation that takes place because of incarceration and because of homelessness, they would be horrified.

—Michael Faenza, President,
National Mental Health Association,
Psychiatric Times, **September 1998**

The name was innocent, even beautiful: St. Mary of Bethlehem. It was first an English priory, founded in the thirteenth century, and taken over by the City of London in 1547. Somewhere in the march of years it became an early asylum for the insane; and its name, shortened to "Bethlem" or "Bedlam," became synonymous with chaotic madness. By the time its long history ended in 1948 and the original Bedlam closed its doors, American mental hospitals were housing roughly half a million patients. Soon after, in the 1960s and 1970s, wholesale release of these patients led to growing signs of Bedlam on the loose.

One early and apparently harmless representative of this unfettered Bedlam was observed by Jonathan Leonard's father, a journalist who commuted into New York City in the 1960s, and who customarily walked from Grand Central Station in mid-Manhattan to the Time-Life Building some blocks uptown. On his way he would often pass a person he called "the Viking." This was not a real Viking, of course, just a man bearing a spear and dressed in Viking garb. The Viking said nothing; nor did he show any sign of working for an advertising firm or doing anything official. He was just there. He was probably deranged. But he was not doing any obvious harm, and so the powers that be let him stand his watch.

As the years passed the Viking grew older, and then one day he vanished. We never learned what happened. Maybe he died; or the police carted him off; or the different drummer in his head told him to go elsewhere; or maybe he got better and the drummer in his head departed. Whatever the case, many thousands of New Yorkers remember him fondly as a symbol of the city's mind-bending diversity and local color.

But the Viking also stood for psychiatry's troubled state and for the multitudes of people with severe psychiatric ills then being expelled from the nation's mental hospitals. In 1955 the number of public mental hospital inpatients in the United States was around 558,000. By 1965 it had declined to 475,000; by 1975 it had dropped to 193,000; and it kept falling—to less than 60,000 at the end of the millennium. If the general population's increase (roughly 72 percent) over this time span is considered, it appears that today's mental hospital inpatient population would exceed 950,000 (more than fifteen times the current figure) had not this powerful urge to empty the mental hospitals intervened.

Of course, even in their heyday the mental hospitals never housed more than a fraction of the millions of people with schizophrenia, bipolar disorder, major depression, and other ills. What they did was take in people who were actively psychotic (unable to relate to reality), later discharging those who responded to treatment, got well on their own, or for whatever reason seemed marginally able to resume their outside lives with whatever assistance was available. This meant that besides housing patients who were psychotic chronically (most of the time), the hospitals also served a much larger population with one-time or periodically recurring attacks of acute illness. This admission and discharge pattern fit well with prevailing patterns of unmedicated psychosis that sometimes persisted indefinitely, sometimes occurred just once, and sometimes recurred periodically like a series of volcanic eruptions.

In hindsight, the mental hospitals' performance was far from perfect. Typically, the patient-to-staff ratio was overwhelming; the care provided was marginal and custodial; many patients were simply warehoused and forgotten; others were admitted when they should not have been, or were kept confined when they should have been discharged. Given the system's size and society's limited capacity to deal with mental illness at the time, it could hardly have been otherwise. Even so, voluntarily or involuntarily, the state mental hospitals gave large numbers of severely troubled people a place to go.

That's what has changed. Today there is nowhere to go. For all practical purposes the place of last resort has disappeared. And while some may hail this demise of the state mental hospital as social progress, that progress has created problems of its own.

To begin with, it has created legal problems. For the purpose of emptying out the mental hospitals and keeping them empty, a welter of laws and court decisions made it illegal to commit psychiatric patients to hospitals against their will unless they posed an immediate danger to themselves or others. Time after time, such danger proved hard to demonstrate in court; and many mental patients with severe ailments, especially when actively psychotic, have what psychiatrists call "poor insight" into their condition. They don't think they are ill, or if they are in a manic state they may enjoy their illness. So they see no need to take medication or go to a hospital, and they are not about to sign a paper committing them to any kind of treatment.

I ran into such a case, Jonathan Leonard recalls, when a long-term acquaintance, a young doctor from the Boston area, made an unexpected visit to my home. James Austin, as we shall call him, was then going through a bout of acute psychosis. He had driven down from Boston to his summer home on Cape Cod, but he had trouble returning because his car wouldn't run on the milk he put into the gas tank. He then proceeded to surprise various old Cape friends with his behavior—including a neighbor he had known from childhood, whom he alarmed by taking a butcher knife from her kitchen and plunging it into her dining room table a foot from where she sat.

She called the police. The police picked up James and took him to a local hospital, where he showed no obvious signs of committing mayhem. The hospital released him a short time later, and in due course he arrived at my doorstep, fondling a baseball bat he had brought along for "protection."

At first I thought this was just a social call from an old friend. But the bat made me edgy, and he clearly was not himself. I decided to get him away from my wife and children by taking him for a walk in the surrounding countryside. So we did that, he seeking to draw my attention to his bat, I seeking to ignore it. We discussed all sorts of things, including his situation; and eventually he drifted off to visit other friends.

Following up on his visit, I was surprised to learn from the police and local health authorities that they felt powerless to act. Clearly, they did not consider his recent history or his vaguely threatening behavior toward me cause to intervene. Hence, unlike the social net provided for someone who commits a crime and gets arrested, or who suffers a medical emergency and gets hospitalized, there was no social structure in place to see that James got scooped up and given proper care. Instead, he and his visitees were left to work things out for themselves.

After a good deal of further turmoil, somebody got the name of a caring relative in Boston who was familiar with James's case. This relative drove down, picked him up, and returned him to his customary home territory, medication, and stability. By then his psychotic episode may have run its course, and he was probably ready to go home. But his visit left me with an indelible impression of how odd the laws pertaining to deranged behavior had become—indeed, how application of the law seemed more deranged than the behavior of my friend. I didn't know then that the legal practices I observed had been left over from a campaign to empty mental hospitals; I just knew that what I saw didn't work.

Besides creating legal problems, decanting the mental hospitals also created problems for the general hospitals not specializing in mental illness. Because their costs are high, general hospitals usually subscribe to a policy of "up and out"; nor are they designed to care for mental patients who are energetically divorced from reality. As a result, many mental patients, especially those with the worst problems, have trouble getting in; and if they do get in, legal and other considerations soon eject them. So the general hospital, even one with a well-heeled psychiatric ward, rarely keeps mental patients long enough to plan treatment, provide medicine, and affirm that the medication is taking hold. Instead, like the hospital where James went, general hospitals tend to act like revolving doors that receive mentally disordered patients one moment only to push them out the next. All too often, instead of establishing a rational care pattern, this process breaks the thread of care, thus neutralizing the work of whoever brought the ill person to the hospital in the first place and sometimes making things worse rather than better.

If the hospitals can't or won't have them, where do these people ultimately go? That's an important question, because anyone joining our crusade to reform psychiatry should know what's happening. Obviously, some of the neglected wind up homeless or in jail, and some commit suicide or other acts of violence. But are we talking about a minor social problem or a huge one? That is, how many people have severe mental illness? How many get treated one way or another? How many are neglected? And how many fall completely through the cracks to wind up as suicides, assailants, street people, or jail inmates?

To begin with, how many Americans have severe mental ills? A prime source of consolidated figures on this subject is a 1993 report by the National Advisory Mental Health Council (an arm of the National Institute of Mental Health) that was made to the U.S. Senate Appropriations Committee. The report, published in the *American Journal of Psychiatry*, found that over a one-year period more than 5 million American adults had schizophrenia, schizoaffective disorder, manic-depressive disorder, autism, or a severe form of some other mental ailment such as major depression, panic disorder, or obsessive-compulsive disorder. That's roughly three out of every hundred adults.*

---

*There appears to be no reason to suppose that the overall prevalence of these disorders has changed markedly since 1993. If one simply allows for the recorded 9 percent growth of the U.S. population between 1993 and the present, the current equivalent of the 1993 figure would be slightly less than 6 million adults.

The Council also found that "during a one-year period, approximately 60% of the adult population with severe mental disorders sought outpatient care for those disorders in the health care system, either in the specialty mental health sector or in the general health sector."[1] Perhaps the most important thing about this statement is what it doesn't say. That is, since only about 60 percent were receiving outpatient care for their disorders, roughly 40 percent of those with severe mental ills (about 2 million adults in 1993) were receiving no outpatient care.

That's dramatic. Of course, the statement doesn't cover inpatient care. But the inpatient population of mental hospitals was very small (about 0.08 million) in 1993; inpatient stays there and elsewhere tended to be brief; and any sort of adequate treatment for a discharged person would have required at least some outpatient visits to see how things were going. It is also inconceivable that any adequate course of treatment would have included a yearlong health care "holiday"—because drug prescriptions need renewing, the diseases have ups and downs, the patients' living conditions change, and so a yearlong absence really means that the patient has been lost to treatment. In other words, the conclusion seems inescapable: as of 1993 some 2 million American adults with severe mental disorders were being neglected, receiving only a lame excuse for treatment or none at all. Nor does time seem likely to have eased this problem. On the contrary, recent trends reducing the quality of psychiatric care make it likely that things have gotten worse.

This helps to explain why so many of the severely afflicted commit suicide. That's no small matter, because suicide has become a leading cause of death in the United States. Killing almost as many people as heart disease, suicide has emerged as a major cause of death among the elderly and also as the fifth leading cause of death among Americans under 45 years old. In all, more than 31,000 Americans kill themselves each year.

Some 90 percent of those who die have a diagnosable mental problem or alcohol addiction when they commit suicide. Available data indicate that roughly 10 to 13 percent of all schizophrenics, 10 to 15 percent of everyone hospitalized for depression, and 15 to 17 percent of all manic depressives ultimately commit suicide.

Unfortunately, the mere act of seeing a physician is not necessarily a good preventive measure. Indeed, a large share (up to two-thirds) of all suicide victims see a physician in the month before their death. That's partly because many of the physicians visited are general practitioners ill-prepared to cope with mental ills, and also because many drugs (including many psychiatric drugs) can be fatal if taken in large doses. Of course, the problem is made worse by HMO rules, limited bed space, and legal restrictions on inpatient commitment that make it hard to admit possibly suicidal people to mental hospitals and even harder to keep them there long enough to ensure they are not a real threat to their own safety. All of this is highly relevant, because it appears that many of these suicides occur during times of acute psychosis, and also during times when the suicidal person confronts troubling conditions, such as being locked up in prison or being homeless.

Of course, most of the mentally ill are like James. They may plummet downward; but before they reach some decisive, degrading, or fatal end point their fall is interrupted—usually by parents, relatives, or close personal friends. Even so, it is worth considering how the various end points of neglect are reached, the numbers of people who reach them, and how the numbers and conditions of these people might be changed.

Let's start with homelessness. It's not hard to see how people with severe mental ills can become homeless. The possible scenarios are endless. But if someone free to roam the streets is psychotic or becomes psychotic, and if he (or she, though there are more homeless men than women) outruns the resources, energies, and patience of whoever may be trying to help, and also avoids jail and the police, that person is likely to become homeless. This presumably explains why the streets have become one of the great repositories of our neglected mentally ill population.

It also explains why housing is not the main issue. For at heart the great difficulty confronting most of the mentally ill homeless is not lack of affordable housing but the presence of psychosis. As E. Fuller Torrey indicates in his book *Out of the Shadows: Confronting America's Mental Illness Crisis*, most of the mentally ill homeless became homeless because of their mental illness, rather than vice versa. Among other things, a large share of the mentally ill homeless suffer from schizophrenia, major affective disorder, or some combination of the two—ailments unlikely to have been caused by a preceding homeless state. So it is reasonable to assume that homelessness did not cause their illness; rather, the illness caused them to lose track of their resources and become homeless. Hence, they need treatment before they can have any real chance of effectively running their own homes; and the understandable but misguided notion that simply providing them with homes can end their "homeless" plight is a delusion.

The bizarre tale of Joyce Brown, recounted in *Out of the Shadows*, illustrates this point. A homeless resident of New York City in the 1980s, Brown lived on a steam grate and was noted for unsanitary behavior (urinating or defecating on the sidewalk, in the gutter, and on herself), exploding into obscenities, running unpredictably into traffic, and wearing little clothing despite the cold. One day the mayor of New York, Ed Koch, saw Brown in action. Realizing that she needed treatment, the mayor asked mental health professionals accompanying him to have her hospitalized. Shocked when told they couldn't do so because she presented no immediate danger to herself or others, Koch had his people develop new, less restrictive guidelines for emergency psychiatric hospitalization. His comment at the time: "If the crazies want to sue me, they have every right to sue, and by crazies I'm not talking about the people we're going to be helping. I'm talking about those who say, 'No, you have no right to intervene to help.' "[2]

Joyce Brown was later picked up and involuntarily hospitalized under Koch's guidelines, and the New York Civil Liberties Union took the case to court. After hearing extensive arguments, the judge ordered Brown's release, noting in the process that "the sight of her may improve us. By being an offense to aesthetic senses, she may spur the community to action." Following several rounds of well-publicized litigation, Brown was duly released; whereupon she appeared on *Donahue* and at a Harvard Law School forum claiming she had never been mentally ill and that her homelessness was caused by lack of housing. She was given a place to live, but around then the mental illness at the root of her condition resurfaced. Soon she was back on the streets, where for some time she remained, providing a sad but convincing demonstration that her troubles arose not from a housing problem but from the state of her own mind.

How many mentally ill Americans are homeless at any given time? As of 1996 the U.S. Department of Housing and Urban Development estimated the number of homeless Americans at about 600,000 and the number of homeless people with serious mental illness at around 200,000. These figures agree with a robust group of independent studies suggesting that about a third of the country's homeless people have severe mental illness—usually schizophrenia, major affective disorder, or some combination of the two.

All too often the conditions of their homeless life turn out to be subhuman. Most people would find the realities of urban street life, even if cushioned by a homeless shelter, to be harsh, dangerous, and degrading. But most people don't have severe mental illness, so in theory they might manage. The mentally disordered homeless generally can't. That's why they became homeless in the first place. Even if their plight is not aggravated by alcohol or drug abuse, as it is in many cases, they are still highly vulnerable. So they commonly get robbed, beaten, raped (if they are women), or put on drugs by addicted acquaintances who frequent the streets and shelters. Too disorganized to get or keep money, and uncertain about food, they may be reduced to scavenging garbage cans. They also run high risks of disease, accidents, and death, many times the risks run by ordinary people; and they sometimes fail to obtain even marginal shelter in cold weather, making them highly prone to frostbite and death by freezing.

While few of us are likely to become homeless ourselves, we should remember that a mentally ill homeless person could be anyone. Most severe mental ailments are "equal opportunity" diseases. They don't pick on the impoverished, nor do they exempt the bright. And while they sometimes strike in youth (the high school or college years), they often wait until one graduates from college, or gets to be a doctor like James Austin, or establishes a successful business like Alice Morrisey. Indeed, a large share of the mentally ill homeless have some college education; and at least one mentally ill 1980s New York street-dweller, a woman named Rebecca Smith, who froze to death in her cardboard box at the age of sixty-one, is known to have been valedictorian of her college class.

How long the homeless state may last is hard to tell. For some it is mercifully brief. The rough interlude may even end happily if the psychotic state abates and the afflicted person seeks out friends and treatment. But it can end otherwise. Street life has its perils; suicide is common among the homeless; and street smarts are not a noted strength of the deranged. The homeless period can also end (or be temporarily interrupted) by a trip to the hospital. Or it can end when the homeless person runs afoul of the law and lands in jail.

If anything, the role of jails seems to be increasing. Big city concern about homelessness has been rising, not so much from sympathy with the homeless as from concern about high levels of public messiness, alcoholism, drug addiction, crime, suicide, violence, and bizarre behavior. So big city governments seem increasingly inclined to replace the municipal helping hand with a mailed fist.

Accenting this trend, in November 1999 New York mayor Rudolph Giuliani announced a new policy: Shelters would no longer be safe havens, because able-bodied shelter dwellers who refused community service jobs would be evicted. That policy, which became a political hot potato, was soon overruled by the courts. But it sent a message to the homeless throng from bedlam, those who could not be expected to cope with any job, that big-city patience had eroded, and that regulations designed to police them up and force them out were tightening.

It is unlikely that this trend will spur any mass migration of mentally ill homeless people to the suburbs, because most suburbs are intolerant of bizarre behavior and quick to repress it. Instead, "sweeping" the city streets of mentally ill vagrants and trying to deny them shelter are more likely to cause some movement to urban fringe areas, an upsurge in the numbers of street people arrested, and expansion of the mentally ill population in jails and prisons. This last result seems especially likely because our prison system already serves as the largest single repository for mentally ill people who have hit bottom.

In his 1872 social satire *Erewhon*, Samuel Butler invented a mythical world where criminals were "cured" of their proclivities by dosing them with unpleasant medicines and sticking them with needles. But citizens unfortunate enough to be sick, because they posed a threat to those around them, were tossed in jail. Today, over a century later, we have come to recognize that mental illnesses like schizophrenia, mania, and depression are not the result of voodoo spells or demonic possession but instead are physical disorders of the brain very much as heart disease, arthritis, and diabetes are physical disorders of the body. Nonetheless, lacking a good place to put mentally ill people, our police, judges, and municipalities commonly throw them into jail.

The result is predictable: "Jails Become Nation's New Mental Hospitals" blared the headline of a recent news article summarizing a 1999 report by the U.S. Justice Department. That report, the first comprehensive national analysis of emotionally disturbed people in jails and prisons, estimated the number of mentally ill inmates at approximately 280,000, over 15 percent of the entire U.S. jail and prison population.

This estimate was based on 1996 and 1997 survey responses by a representative sample of jail and prison inmates who were asked whether they had a mental or emotional condition, and whether an emotional or mental problem had ever caused them to stay overnight at a mental hospital, unit, or treatment program. A "yes" answer to either question was read as indicating mental illness. Though no degree of severity was specified, the results appear consistent with other limited prisoner studies that have found 8 to 16 percent of the prisoners surveyed to have a severe mental illness (schizophrenia, bipolar disorder, major depression). The results also seem consistent with reports that large numbers of severely disordered inmates have been a high-profile feature of jail and prison populations ever since the mental hospitals were decanted.

In that case, one might ask, why shouldn't prisons serve perfectly well as impromptu mental hospitals? After all, the mental hospitals became notorious for warehousing mental patients. If that is essentially what prisons do, what's the difference?

This question has two answers. First, nobody is seriously urging revival of the large state mental hospitals, so they provide a poor standard for comparison. And second, prisons generally treat their mentally ill inmates worse than such people are treated in mental hospitals. That's because a prison's purpose is not to prescribe psychiatric medicines and therapy, ensure that sound treatment is received, and assess what's going on. Rather, its purpose is to confine and perhaps reform criminals. To that end it fills the prisoners' days with rigid rules and punishes the disobedient. Typically, those with severe mental disorders are in no position to obey such rules, so they get punished—with or without the medications that they need. If they are put into prolonged solitary confinement (a common fate for trouble-some psychotics), that could be roughly comparable to putting an asth-matic in a smoke-filled room. On the other hand, if they remain in the com-pany of other prisoners, especially if they draw attention to themselves with bizarre behavior, they become ready targets for brutal beatings, rape, or murder.

Of course, the reverse is also true. Serious mental patients are not good for prisons. They create noise, cause messes, break rules, and make a mock-ery of prison discipline. They also present easy targets for the sorts of vio-lence and brutality that the prison is (or should be) trying to discourage. For all of these reasons, as one jail official put it, "the bad and the mad just don't mix."

But if psychotic people don't belong in jail, why are they sent there? Samuel Butler would understand. They are sent there "for their own good," because there is nowhere else for them to go, and also to keep them from harassing, threatening, or harming others. Often the criminal charges against them are hardly more than an excuse. Formal accusations of "loiter-ing," "trespassing," "disorderly conduct," and "petty theft" (often of food or restaurant meals) commonly serve to shoo mentally ill troublemakers and hardship cases off to jail.

More often than one might think, a problem-solving sort of charity may prompt the arrest. As recounted in *Out of the Shadows*, one day police in Madison, Wisconsin, were called to deal with a mentally ill woman yelling in the streets. She didn't show the kind of "dangerous" behavior needed to commit her to the state mental hospital; nor did she show any interest in go-ing to a shelter or taking medicine. But she clearly was creating a nuisance and risked being assaulted in the street; so the police arrested her, charged her with disorderly conduct, and jailed her for her own protection.

In this same vein, a policeman quoted by the *Los Angeles Times* in August 1991 described homeless people with severe mental ills as being found "suffering from malnutrition, with dirt-encrusted skin and hair or bleeding from open wounds. . . . It's really, really pitiful. . . . You get people who are hallucinating, who haven't eaten for days. It's a massive cleanup effort. They get shelter, food, you get them back on their medications. . . . It's crisis intervention."[3]

Even so, the effects are often brief. Jails, generally defined in most states as places housing inmates for relatively shorter periods than prisons, are often just revolving doors—places that provide brief respite without effectively assessing or treating people. All too many provide "crisis intervention" and little else. In the words of Kay Redfield Jamison, a psychiatry professor at the Johns Hopkins University medical school: "Incarceration of the mentally ill is a disastrous, horrible social issue. . . . There is something fundamentally broken in the system that covers both hospitals and jails."[4]

Besides targeting jails, Jamison was probably also targeting prisons, where the *Erewhon* mentality persists. There, however, the situation tends to be more dangerous. Typically, mentally ill prison inmates are in more trouble than mentally ill jail inmates because they have committed greater offenses, are surrounded by more dangerous people, and are incarcerated for a greater length of time. But the fact remains: their incarceration is bad for themselves, for the other prisoners, and for the prison system. Often they are in prison mainly because they have done something that demands confinement, and because there really is nowhere else for them to go.

One sad fact that needs to be recognized here is that actively psychotic people can be violent. They can't cope with reality (that is the definition of psychotic). So their view of reality sometimes clashes with the truth, and when that happens they react.

I had a patient, Allan Hobson recalls, a young electronics worker suffering from schizophrenia, who provided a good example of this behavior. The time was the early 1960s, the place the Massachusetts Mental Health Center, where I was a new psychiatric resident. Freudian psychoanalysis reigned supreme at Mass Mental in those days, and I was advised not to medicate this patient because that would block his true feelings and prevent him from entering into a sound psychoanalytic relationship with me.

He did get medicated, however, because he had severe hallucinations about being in a war zone; and one day, when my supervisor approached him, he mistook this supervisor for the enemy of his visions and beat him up. Psychoanalysis or no, that caused him to get medicated with an early neuroleptic, Thorazine (chlorpromazine), which settled him down enough in a few days so that he could discuss his problems calmly in my office. Besides making me a little more skeptical about psychoanalysis, the incident served as a good example of how delusions can cause psychotic people to become violent.

As this suggests, in many cases the violent person doesn't plan to cause mayhem. But if a woman imagines she is a store clerk, for example, she may punch a real store employee who tries to keep her from rearranging items on the shelves. Or someone who believes he's Jesus Christ may try to break into a locked church, or may assault someone he identifies as Judas or the Devil.

Perhaps mainly for this reason, a wide range of studies by different investigators have shown that people with severe mental disorders are prone to violence—especially against family members, and especially after they have failed or refused to take their medications. In general, three risk factors make violence by the severely disordered far more likely than it would be otherwise: a past history of violence, existence of a drug or alcohol problem, and failure to take medications.

Sometimes madness can prompt murder. "New York Nightmare Kills a Dreamer" shouted one front-page *New York Times* headline about a January 1999 subway murder. The victim, Kendra Webdale, was a pretty young blonde woman in love with New York City. Her accused killer, Andrew Goldstein, was a withdrawn loner with a long history of schizophrenia. Goldstein did not have a violent past or any obvious alcohol or drug addiction, but he had gone off his medication. Some time after that happened, according to the *Times*, he entered a subway station, went up to Ms. Webdale, apparently asked for the time of day, and then pushed her into the path of an oncoming train, where she was struck and killed.

Despite the horror, human interest, and notoriety of this case, we should not be too surprised. One consequence of our failure to deal with bedlam on the loose is that crazy subway assaults happen. Indeed, a research article published in the *General Archives of Psychiatry* in 1992 by Daniel Martell and Park Deitz suggests they are fairly common. The article, entitled "Mentally Disordered Offenders Who Push or Attempt to Push Victims onto Subway Tracks in New York City," describes a study of 36 people who had previously tried to kill others in this manner. In one case robbery motivated the assault; but in all of the other 35 cases the investigators found evidence that the attacker was psychotic.

In contrast to bizarre murders, which get intense media coverage, most lesser acts of violence by mentally ill people go unreported. Even so, police have become increasingly accustomed to dealing with the emotionally disturbed. In Memphis, Tennessee, police with special mental health training are called in when an unstable person is involved. And in New York City the police receive sixteen hours of training on how to deal with emotionally disturbed people—which is just as well, because in 1998 the New York City Police Department fielded 60,000 emergency (911) calls dealing with such people.

Given this volume of mandatory-response calls, it's not surprising that things sometimes end badly. In September 1999 the New York City police were called in to deal with Gidone Busch, a man who reportedly believed he was directed by God to save drug addicts and exotic dancers. Busch was said to be menacing children in a Jewish section of Brooklyn. When police arrived on the scene, Busch attacked the officers with a claw hammer. The police shot him dead with twelve bullets.

Some have asked whether the police did the right thing in this and other situations, and whether they should have been in charge, because they are not mental health workers. But in most cases they have no choice. Their job is to quell public disturbances. Therefore, if mentally ill people are creating such disturbances and nobody else is poised to intervene, the police must do so, whether they are well-suited to intervene or not.

A more troubling question is whether the police can make much of a dent. Obviously, most incidents are resolved without killing the perpetrator. So what can the police do then? They can simply leave, setting things up for a recurrence of the problem. Or they can take the disturbed person to a jail or hospital, which all too often acts like a revolving door.

In some cases, of course, police may get the afflicted person back into effective treatment and into the hands of friends, relatives, or parents who will help maintain it. That solves the immediate problem. But as we have seen, most people with severe mental illness aren't actively psychotic all the time. And we are dealing with roughly 2 million people who have severe but untreated mental ills. So it's a little like punching pillows. No sooner do you control the problem in one place than it bursts out somewhere else.

The same can be said of most measures designed to help the homeless or treat people in prison. Many of these measures are worthwhile, but they can't solve the problem of the mentally ill homeless or imprisoned, because those who are homeless now (roughly 200,000) or imprisoned now (some 280,000) are only part of the problem—the currently active part. The rest of the problem, posed by the vastly greater multitude with marginally treated or untreated disorders, stands ready to become active as members of this multitude experience psychosis and fall through the social cracks, sometimes to a point where they become violent, homeless, or imprisoned. In other words, our *Erewhon*-like ways of handling severe mental ills simply don't work. Bedlam on the loose will remain a free-running Bedlam until we provide severe mental cases with proper treatment, or unless we imprison so many (900,000 or so) that we keep nearly all active psychotics warehoused—as they were in the old asylums.

Those tempted by this latter option should note that imprisoning 900,000 people would be costly. Indeed, at today's prices (more than $30,000 in capital and operating expenses per year for each inmate) it would cost over $27 billion yearly. Even imprisoning 280,000 costs over $8 billion yearly, and this figure does not include the expense of police, lawyers, and courts, or the social cost of dealing with 2 million untreated people who are actually or potentially psychotic, including 200,000 homeless people and innumerable others who are capable of acts that are antisocial and sometimes violent. All in all, it seems clear that the only winners in such neglect-ridden circumstances are the HMOs and insurance companies who don't have to pay for the neglect.

If we don't like the current picture (we shouldn't, because it's probably more costly, dangerous, and cruel than the old asylum system), and if we don't want to bring back the asylums, the only answer is to effectively treat those who are neglected. And since we have both the medicines and the knowledge needed to tackle this large task, the technical side looks bright.

But what of the human side? Clearly, we have no effective health structure assigned to receive, treat, and keep tabs on mental patients; and most of the limited structures we do have are working in a fragmented, partial, or poorly coordinated fashion. Indeed, instead of looking like smoothly coordinated health services, our ramshackle system looks far too much like the croquet game in *Alice in Wonderland*, where the flamingoes, soldiers, and hedgehogs ordered to serve as mallets, wickets, and croquet balls spend a lot of time getting in the way and doing the wrong thing.

Given this general picture, anyone can see that it's not enough to have psychiatrists and psychologists waiting in the wings to receive patients, or for sound support groups to emerge that try to help the ill, or for parents and relatives of mentally ill people to take up slack by coordinating care and monitoring treatment. Of course, all of these things are vitally important; but they are not enough, because despite them multitudes of severely ill people fall through the cracks with damaging and sometimes horrible results. What we really need to do, therefore, is to gear up for a scientific, medical, and political crusade that will do two things: blend our growing biomedical knowledge with sound therapy and create an effective mental health structure that will give comprehensive care to the severely ill who are neglected.

What role should psychiatry play in all this? Normally that answer would be automatic, because psychiatry's duty is to confront this challenge; psychiatrists are the only doctors specifically trained to deal with mental patients; and today's neglected multitudes are in fact psychiatry's lost charges. But psychiatry, like its charges, has fallen on hard times. It has developed something resembling a split personality disorder, as a result of which it has lost a good deal of its former prestige and clout. And indeed, psychiatry's fall from grace has a lot to do with our currently disgraceful treatment of the mentally ill. Therefore, it is important for everyone involved to understand psychiatry's past, where it stands right now, why it has lost ground, and how it might recover. That is especially important because, as shall be seen, psychiatry with all its troubles still has the best chance of leading a strong charge toward proper mental health care and sound treatment.

3

Psychiatry's Rise

*When the psychiatrist Herbert Meltzer did his post-M.D.
training (his residency) in the mid–1960s, the Massachusetts
Mental Health Center was at the epicenter of American
psychiatry. "It was dominated by psychoanalysts," he recalls,
"who were totally committed to psychodynamic
[psychoanalytic] psychiatry. Even their approach to
schizophrenia was psychoanalytic."*

—**Edward Shorter,**
"Prozac v. Freud: Medicine Wins,"
Current, **January 1999**

Nobody worried about what psychiatry was doing in the eighteenth century, because there was no psychiatry. But in those days anyone could see that many of the insane were treated badly, and that many had nowhere they could go. First in Europe and later in America, this problem gave rise to a humanitarian concern that found its voice in the early nineteenth century. Out of this emerged a major asylum-building campaign and the first psychiatrists, the latter being directors of the asylums.

These early psychiatrists did not plan to warehouse the insane but rather to provide effective treatment. Chances are they could not have done much anyway, given the primitive state of knowledge in those days. But they never got a chance, because the asylums proved vastly popular. So their good intentions were washed away in an overwhelming flood of patients, and by the 1860s psychiatry's main role was to serve as a gatekeeper for patients receiving care that was mostly custodial.

Around this time, psychiatry had a serious run-in with neurology. In a recent conversation E. Fuller Torrey, an expert on this period, summarized the confrontation as follows:

> Neurology came out of the Civil War, inspired by examination of war-related head injuries; and when it organized in the late 1860s and early 1870s it tried to take over psychiatry. It was a fascinating fight, a bitter battle in the professional journals and the *New York Herald Tribune* using some of the most vituperative language I have ever seen in print.
>
> Basically, what the neurologists said was "Both of us are dealing with the same brain. But you guys off in your asylums are hopeless. You're not doing any research, your asylums have gone downhill, and we demand a state investigation." They got the investigation; they carried on the fight, mostly in New York State, a little in Massachusetts; and they made a very strong attempt to take over psychiatry.
>
> In the end they failed, but not by much. They failed because the asylums were then growing very fast. And when push came to shove and they were asked what should be happening, they didn't have really good answers. They didn't have an alternative plan other than to say "You ought to take the jobs away from those guys because they don't know what they're doing and give the jobs to us." And the state people correctly said "You have something of a vested interest here, don't you, so why should we believe you?" And neurology then went off and became very much what it is today, taking care of all the brain diseases not presenting psychiatric symptoms.

This early clash, in some ways reminiscent of what is happening today, led to a situation where psychiatrists were largely limited to working in their own state mental hospitals. But psychiatry was sustained because the hospitals kept growing. Between 1903 and 1933, the total inpatient population of the state mental hospitals rose from 143,000 to 366,000. Most of these hospitals had over a thousand beds, and working there was depressing. As Edward Shorter notes in *A History of Psychiatry: From the Era of the Asy-*

lum to the Age of Prozac: "From the viewpoint of physicians who had to practice in them, asylum psychiatry counted scarcely as a branch of medicine at all. One could cure nothing. There was little scientific understanding of mental illness. And one lived in rustication far from the medical centers with their state-of-the-art labs and great libraries. Younger, often idealistic psychiatrists bridled at this sterile incarceration and sought alternatives."[1]

One of the best alternatives that materialized in the early 1900s was smaller urban hospital facilities near city libraries and teaching centers. These could assess mentally ill city people far better than the police or jailers who typically screened them, and so could act as clearinghouses for deciding which patients needed to be sent to the large hospitals. Lacking the crowd pressure of the large hospitals, they could also provide therapy and individual attention for selected patients. Since they could thus provide better care, they could usually have their pick of mental patients. And they could also draw on the literature, students, teachers, and researchers of nearby city libraries, medical schools, and universities. Thus these smaller facilities found themselves in a good position to conduct pioneering mental health research—an especially appealing incentive for young doctors seeking alternatives to the asylums—and so they became bright beacons for psychiatry and gathering points for local talent.

One of these urban centers, now known as Boston's Massachusetts Mental Health Center, was established as a subordinate part (the "Psychopathic Department") of the Boston State Hospital in 1912. We shall focus on this facility because one of us (Allan Hobson) has spent much of his career there; because its history mirrors much of psychiatry's past; and because it now faces a crisis similar to that confronting psychiatry as a whole.

Initially, besides screening mental patients, the Psychopathic Department forged strong ties with the nearby Harvard Medical School and set up a first-rate brain disease research laboratory. By 1920, when it became an independent state hospital (the Boston Psychopathic Hospital), its research had advanced our understanding of delirium tremens and had made

FIGURE 3.1
The Massachusetts
Mental Health
Center was located at
74 Fernwood Road
in Boston in 1912 to
assure easy access by
the city's many
public patients as
well as strong
scientific interaction
with the
Harvard Medical
School.

important inroads against the ravages of neurosyphilis, a disease then responsible for 10 percent of all U.S. mental hospital admissions.

Between the world wars the Boston Psychopathic Hospital flourished.
Encouraged by state health authorities to be creative, and nourished by
Harvard as well as by Boston's medical establishment, it developed a top-
flight reputation and attracted leading researchers, teachers, and medical
school graduates seeking psychiatric residencies.

The main thing limiting its appeal was the same thing limiting the appeal
of psychiatry generally: its medical approach. That approach saw mental
ills as being caused by organic disorders of the brain. And while this position is perfectly compatible with rapid progress against mental ills in today's world, in those days the store of medical knowledge was small, drugs

were primitive, and human understanding of the brain was dim. Therefore, the prospects for medical progress were limited, and the main employers of psychiatrists remained the large and forbidding state hospitals. So it is hardly surprising that most medical school graduates looking for desirable specialties chose other fields.

What changed all this and turned U.S. psychiatry on its head was Freudian psychoanalysis. Besides offering hope, psychoanalysis offered an appealing alternative to the mental hospitals. And while it often explained mental ills in ways that were bizarre, erroneous, misdirected, confusing, and sometimes harmful, its theories were based on keen observations of human behavior that seemed at least as good as any others available at the time. So it is worth recalling who Sigmund Freud was and what he did.

Freud was born in 1856 and spent most of his life in Vienna. Trained in the medicine, physiology, and neuroscience of his day, he pursued basic scientific research in his twenties and thirties before becoming fascinated with psychology. In 1895, after noting that many patients who consulted him for neurologic symptoms had no signs of brain disease, he proposed developing a "scientific" psychology, one founded upon existing neuroscience knowledge; but he abandoned this project because too little was known about the brain.

Freud was also deeply attracted to psychology through hypnosis. He had observed hypnosis at Jean Martin Charcot's clinic in Paris, and his friend Josef Breuer had demonstrated that hypnosis could apparently cure cases of "hysteria" by causing a hysterical patient to recall forgotten memories somehow tied to the disorder. But Freud did not pursue hypnosis. Instead he embraced the idea that forgotten or "repressed" memories—typically childhood memories related in his view to sexual urges—were responsible for mental ills, and that rediscovering and examining these memories could cure the patient.

In essence, Freud's psychoanalysis sought to recover and assess these memories using a variety of methods. One method was dream interpretation, in which the recalled dream was decoded on the assumption that it represented repressed unconscious wishes that had been disguised in a symbolic manner to protect consciousness and thus preserve sleep. Another was "free association," in which patients were encouraged to let their minds touch repressed ideas by saying anything that came into their heads. And still another was "transference," whereby the psychiatrist came to represent one of the patient's parents and the patient came to work through childhood feelings and attitudes relating to that parent.

These methods took considerable time. For instance, one can hardly imagine "transference" taking place unless the patient has had time to develop a close relationship with the analyst. So it's not surprising to find that the analyst almost had to become a temporary member of the patient's family; or that multiple one-hour sessions (commonly up to five a week) were recommended for extended periods; or that the costs of psychoanalysis demanded a fat wallet.

Another problem: The benefits of psychoanalysis were hard to prove. Not only are mental ailments quirky, but Freud's theory enjoyed no firm foundation. As he himself had shown, the rudimentary neuroscience of his day could not provide any such foundation. So, like all other psychologists of his time, he simply depended on observations of mental illness and human behavior to develop what today are called "top-down" theories about how the mind operates and how mental illness might be cured. He did have one advantage: his experience could be used to ensure that his theories did not conflict with prevailing neuroscience concepts. But aside from that he really had no choice except to engage in educated guesswork, which is what he did.

Of course, around 1900 nobody else was doing any better. Indeed, educated guesswork in a nice office seemed like a big improvement over what the mental hospitals were doing. And clearly, Freud's approach established a caring relationship. So it looked better than competing office methods like electrotherapy that were likewise unproven and so impersonal that they might provide no caring relationship at all. Thus, it is not surprising to find that Freud enjoyed moderate success in Vienna, that his theory competed effectively with those of other psychologists, and that he was able to surround himself with a small group of disciples.

In 1909 Freud visited the United States with Carl Jung to attend a meeting at Clark University that was also attended by William James, the father of American psychology. James was quite skeptical of Freud's theory because it could not be proved or disproved. But that did not prevent Freud's visit from inspiring a small psychoanalytic movement in the U.S. that took root, established itself, and emerged from obscurity in the 1920s.

Of course, psychoanalysts didn't need to be doctors, because they used no drugs or medical methods. In theory, English teachers could do psychoanalysis every bit as well as doctors, maybe better. For in fact psychoanalysis was just an elegant form of applied psychology in which the analyst tried to help the patient with "talk therapy." All the bizarre sex-related concepts, things like the Oedipus complex, or the odd theories about dream analysis and the subconscious, were really just psychological ideas that Freud had invented or borrowed from others. None was proven, and none had clear medical implications or required an M.D. to apply.

But the ills psychoanalysts treated were the same, though generally less severe, as those psychiatrists were treating. That is, they were ills without any obvious organic cause that affected the patient's conscious mind. What's more, besides sharing a common group of ills, U.S. psychoanalysis and psychiatry had a lot to offer one another. Psychiatry could offer the newcomer entry into a large and well-established branch of American medicine that was treating over half a million patients. And psychoanalysis could offer something even better—a way out of the dead-end mental hospitals and into comfy, fashionable private quarters. So it's not surprising that the two fields got together.

Indeed, "got together" is an understatement. From 1924 onward, the American Psychiatric Association and the American Psychoanalytical Association held their annual meetings at the same time in the same city. And in 1938 the latter, charged with setting training rules, required that all candidates for training in analysis needed to have completed at least one year of a psychiatric residency. In other words, anyone who wanted to be trained as an analyst had to be an M.D. en route to a psychiatric specialty.

Psychoanalysis got the clout to do this from events in Europe. Not all European analysts were Jewish, though Freud and many of the most distinguished were. As a result, the rising tide of fascism in the 1930s drove many of the best-known analysts from Central Europe to America. (Freud himself emigrated to England in 1938, shortly before his death.) These European emigrés were few in number, probably less than 50; but their fame and zeal lent American analysis prestige and transformed it into a beacon capable of attracting large numbers of aspiring young psychiatrists. By the end of the 1930s, psychoanalysis was fast becoming part of general psychiatric practice in this country; and indeed, the overlap between the two was such that by 1944 roughly 70 percent of all U.S. analysts had qualified in psychiatry.

The trend did not stop there, however. Beginning in the 1940s, psychoanalysis began taking over the nation's top psychiatry chairs and university departments, and by the mid-1950s nearly every prestigious psychiatry chair in the country was occupied by an analyst. Within psychiatry, analysts wrote the textbooks, ran the journals, called the shots. For most of the 1950s and 1960s, the president of the American Psychiatric Association was an analyst or someone highly sympathetic to the movement. And while most psychiatrists were not actually analysts themselves (there were not enough institute training slots available), nearly all were oriented toward analysis and steeped in Freudian thought.

The Massachusetts Mental Health Center came to play a leading role in this transformation of psychiatry. From its start in the 1940s, Mass Mental had taken the standard medical approach to mental ills, treating them as organic brain diseases. Harry Solomon, who became its director in 1943, was a syphilis expert. In those days he could be found in the outpatient clinic doing spinal taps and even tapping the fourth ventricle (one of the spaces within the brain) to get fluid from there, because so many patients had syphilis and one could find the responsible microbe (a spirochete called *Treponema pallidum*) in their brain and spinal fluids. But then penicillin came along, which vastly reduced the syphilis problem; and then psychoanalysis arrived in force; and by the 1950s Dr. Solomon was wearing street clothes and talking to patients about their mothers.

Mothers were a key issue in those days. The main Freudian idea was that mental ills were caused by repressed childhood memories. These memories were repressed because they recalled unacceptable natural desires (like the Oedipus complex, in which the young boy supposedly wants to kill his father and marry his mother) or because they recalled taboo responses to parental mistreatment or rejection.

Freud himself had tended to shy away from the psychoses—presumably because the likely gains from talk therapy were pretty limited if the patient couldn't cope with reality and therefore wouldn't talk coherently or couldn't relate to the analyst. But limited efforts were made in the United States to deal with the psychoses through analysis, and these received a major boost from an emigré German analyst, Frieda Fromm-Reichmann, who wrote extensively in the latter 1940s about schizophrenia being caused by childhood rejection, especially rejection of the child by its mother. Armed with this cold "schizophrenogenic" mother concept, and with other analytic explanations for major ills like mania and depression, increasing numbers of analysts felt themselves equipped to deal with the multitudes of seriously ill patients languishing in the nation's mental hospitals. As a result, in 1958 the American Psychoanalytic Association established a program for teaching these doctrines to the mental hospitals.

But the Massachusetts Mental Health Center didn't need to be taught these doctrines, because it already knew them. It was on the cutting edge of the new wave. By then it had 300 beds, cared for over a thousand outpatients, and received a far greater share of the Massachusetts state budget than any of the state's eight other mental hospitals—perhaps as much as all of them combined. Indeed, it had such eminence, such effective ties to Harvard, and such a close relationship with the state mental health establishment that in 1958 its director (Harry Solomon) traded jobs with Jack Ewalt, the Massachusetts Commissioner of Mental Health, Solomon becoming the new state mental health commissioner and Ewalt becoming director of Mass Mental. In light of such events, it's not surprising that Mass Mental stood out as a leading center. Clearly, it was at least as well equipped as any facility, far better than most, to psychoanalyze severely ill mental patients—which is what it did.

Another reason for Mass Mental's popularity was the charisma of its clinical director, Elvin Semrad, whose skill in communicating with psychotic patients became legendary—to a point where students joked that there should be a unit for measuring empathy called the "semrad." Although he did not try to advance psychoanalytic theories directly, Semrad had been trained as a psychoanalyst, and observers naturally associated his interviewing talent with psychoanalytic doctrine. As a result, many of Semrad's best students rushed off to the psychoanalytic institutes hoping to emerge with Semrad's abilities. Although they were mostly disappointed, because Semrad's personal qualities were responsible, that did not detract from Semrad's talent.

For students, all this gave Mass Mental vast appeal. Nobody had proved that psychoanalysis really worked, or that it reduced psychosis, but nobody had proved that it didn't. Indeed, it had a promising air of mystery about it. So there seemed hope for millions of the mentally ill. The drab image of the mental hospital and its sorry medicine had been replaced by caring relationships and hopes of cure in friendly settings. What's more, by then most psychoanalysts were themselves medically educated psychiatrists who earnestly believed that what they did was probably beneficial and clearly better than what had gone before. Small wonder, then, that students coming out of medical school were swept up in the mounting wave of enthusiasm, or that the top graduates of Harvard and other leading medical schools eagerly competed for residency slots at Mass Mental.

This enthusiasm for psychotherapy as the treatment of choice for psychosis was not dampened by the emergence of drugs that were really effective against schizophrenia and other major ills. The first of these, the neuroleptic Thorazine, which became available in 1955, was generally scorned at Mass Mental. As we have seen, residents were told not to use it; and Semrad insisted that giving a patient this or one of the other emerging antipsychotic phenothiazines would separate the head from the heart. So instead, psychiatry was divorced from medicine. Mass Mental tried to make everything as much like a friendly home as possible. Nothing was locked up. And although there were seclusion rooms for temporarily rambunctious patients, every effort was made to avoid confinement and to act in a quasi-parental fashion.

Of course, analyzing psychotic patients in this cozy unlocked setting looks odd now. And indeed, things could get bizarre. Once, Allan Hobson remembers, I found myself trying to interview a patient who had left the main building and taken refuge under a nearby car. I also recall a tall, handsome psychoanalyst from the suburbs, who would come to Mass Mental once a week to conduct a sit-down direct analysis with a totally psychotic female patient, translating everything she said into Oedipal language in the presence of medical students. In hindsight it was quite remarkable.

Another striking thing is that if you asked questions about what was being done, they always referred you to your parents. They'd say "Do you ask that question because you're angry at your father?" In 1960 I was in a first-year seminar on the history of psychiatry taught by Jack Ewalt, the director. One day, when I was giving a report on William James and John Dewey, he said "You know, it sounds like you think science makes progress. Do you really believe that?" Ewalt himself tended to think that science raised more questions than it answered. He was quite pro-science, but he wasn't a scientist. So I said, "Of course I believe it. Don't you?" adding that the importance of science was so clear that it was a little silly to ask such a question. His response to this mild impertinence was "Do you think you're talking to your father?" My retort was "Absolutely not. My father would never ask a stupid question like that." At which point, to Ewalt's credit, he became my fierce ally and defender.

In time I came to see that this exchange and its result revealed something important—not just about Mass Mental but also about the relationship Jack Ewalt and certain other psychoanalysts came to have with science. For by utilizing the resources of the newly founded National Institute of Mental Health, the scientific imprimatur of the Harvard Medical School, and the talent of his young residents, Ewalt in fact became a vigorous promoter of the science he seemed to question. Among the students attending Ewalt's seminar series were Joseph Schildkraut, who later developed the first neuro-biologic theory of depression; Paul Wender, who proposed a brain basis for attention deficit disorder; and Eric Kandel, who won the Nobel Prize in the year 2000 for his work on the neurobiology of learning and memory.

4

Psychiatry's Downfall

*Psychiatry is an emperor standing naked in his new clothes.
It has worked and striven for 70 years to become an emperor,
a full brother with the other medical specialties. And now it
stands there resplendent in its finery. But it does not have any
clothes on; and even worse, nobody has told it so.*

—E. Fuller Torrey,
The Death of Psychiatry, 1974

The new wave of enthusiasm for psychoanalysis did little to harm the large and forbidding mental hospitals directly. But the new wave did other things. To begin with, as most psychiatrists became analysts themselves or analytically oriented, they lost touch with their old haunts. All the prestige, all the status, all the money became concentrated on office-based psychiatry or places like Mass Mental, leaving anyone working in the large, mostly rural mental hospitals in a strictly second-rate position. Thus, these bastions of psychiatry's old medical way of life no longer provided the main base for the profession. Psychiatry had gone on to better things. So there was scant professional need to defend these fortresses, and as time passed that need became less and less.

More damaging to the old medical side of psychiatry's split personality was growing confusion about the nature or even the existence of mental ills.

For most Freudians spurned textbooks, white coats, and medical science. They believed that mental ills were caused by repressed childhood experiences. And since most such experiences didn't seem to fall into neat categories, there was no reason why mental problems should do so either. Thus, the Freudians had no use for the old medically based classification system that tried to separate mental illnesses into various types, starting with those that were reasoning disorders versus those that were "affective" (emotional) disorders, or that tried to distinguish illnesses according to the severity and nature of the symptoms they produced. Instead, the Freudians pushed the idea that everybody, including normal people, had some mental problems that could be resolved through analysis. That's why analysis was recommended for analysts in training, all psychiatric residents, and practically anyone else who could afford it. And that's why people with chronic psychosis who were confined in mental hospitals were deemed to have troubles that differed from those of normal people only in degree rather than in kind.

Instead of being dismissed as bizarre, these ideas won out. A widely quoted Midtown Manhattan study of the mid-1960s found that 80 percent of the population surveyed was psychiatrically impaired, with only 18.5 percent being well and 23 percent being severely ill enough to require treatment. Eight years later the noted libertarian psychiatrist Thomas Szasz could write in criticism: "It is no exaggeration to say that life itself is now viewed as an illness that begins with conception and ends with death, requiring at every step along the way, the skillful assistance of physicians and especially mental health professionals."[1]

From here it was only one step to thinking that the mental hospitals, by neglecting patients in an impersonal setting, were creating more mental problems than they solved, and that everybody would be better off without them. It was amazing, as this last idea emerged, to see how far psychiatry's pendulum had swung from the "mindless brain" side of its personality to the "brainless mind" side, from the purely biomedical side that saw mental ills as organic brain disorders best treated in mental hospitals to the purely psychotherapeutic side that saw these ills as psychological mind problems arising from the patient's past, and that saw confinement in a mental hospital as likely to do more harm than good.

With this idea in place, losing faith in the large state mental hospitals was easy. For they really didn't cure their patients. They really were mostly holding tanks. They really did neglect a lot of people. And many of them really were poorly managed. Beyond that, few psychiatrists worked in those places any more. Certainly the leaders of psychiatry did not. But the key point is not that psychiatry was self-serving. The key point is that by the 1960s most psychiatrists believed in Freud's precepts. So they really did hope that psychoanalysis might cure severe ills like schizophrenia, while doubting that medical treatments in mental hospitals ever would. This explains why the psychiatrists at Mass Mental downplayed the phenothiazines, kept everything unlocked, tried to minimize confinement, and tried to create a cozy homelike setting. However foolish their actions may look in hindsight, these people clearly believed in what they did.

But psychiatrists with these beliefs and no personal stake in the mental hospitals couldn't be expected to defend them. And if they wouldn't defend them, who would? The answer was almost no one. The psychiatrists who worked there might, but they had become the least respected part of the profession. Inmates with severe mental ills, even if they wanted to defend their hospitals, couldn't do so because they lacked credibility. Neither could former inmates who had recovered, most of whom were trying to avoid drawing attention to their past illness. And neither could the families of inmates, who also feared being marked by the disease. That was especially likely in those days, because Freudian doctrine blamed families of the mentally ill for creating the illness, and the stigma associated with insanity was much greater then than it is now. So for both reasons, few of those touched directly or indirectly by mental ills felt comfortable talking about them, much less joining advocacy groups.

That was bad news for the mental hospitals. For the confused body of Freudian ideas about the nature of mental ills acted like an overheated tropical sea. It did not destroy the mental hospitals directly. But it produced rising updrafts of opinion over a large area–updrafts that gained strength, gathered together into a storm of criticism, and blew the mental hospitals away.

The first outriders of this storm could be found in the works of Ronald Laing and Thomas Szasz. Laing, a British psychiatrist, asserted that the mentally ill were no more insane than other people. And Szasz, the libertarian psychiatrist quoted above, claimed that mental illness was a social myth—a myth used to deprive eccentrics and nonconformists of their freedom.

Like strange clouds on the edge of an approaching hurricane, these views gave every sign of coming from the lunatic fringe. But the time and public sentiments were right. So they were warmly embraced, their authors became highly influential, and Laing emerged as a noted U.S. counterculture hero.

One of us (Allan Hobson) went to hear Ronald Laing speak on two occasions. The first time, in the 1960s, he was speaking at Tufts University. Lots of people from the Massachusetts Mental Health Center went over there, because we had read his books on how all you need to do is give the mentally ill a nice home and they straighten right up. He was then living with eight schizophrenics at a house in London.

We went to the Tufts gym to hear him and could hardly get in. The place was packed. It was like a rock event. It really was. People thought everything was going to change—everything. The guy had a world following. He sat on a bentwood chair, lit by a 50-watt bulb, with everything else in darkness. I was hanging off a bannister somewhere. People were hushed, taking in every word he spoke. It was like listening to the Dalai Lama.

Eventually, long after he was famous, he went to India, more or less followed the Timothy Leary path, took a lot of drugs himself, and got burnt out. Somewhere in the latter 1970s he gave a talk at our gym at Mass Mental. By then people were already beginning to say "Wait a minute!" Certainly they could see that he had ruined himself, because he seemed vacuous and disconnected. This didn't disprove his ideas, which by then were losing ground for other reasons, but it spelled the end of his dramatic career in a way that was very sad.

But that was still in the future. In the 1960s, as the storm advanced, anyone could see that the prominent sociologists and others who took up
these ideas and ran with them—blaming the mental health establishment
for improperly "labeling" people as insane—were no longer on the fringe.
Nor was Ken Kesey's popular 1962 novel *One Flew over the Cuckoo's Nest,*
which condemned the asylums and gave rise to a film that later won several
Academy Awards. Nor were other books like Erving Goffmann's *Asylums*
(1961), which blamed the mental hospitals for causing the very conditions
they were supposed to cure. And neither was the work of David L. Rosen-
han, a Stanford psychology professor who sent sane volunteers into mental
hospitals to demonstrate that they would be admitted and misdiagnosed
for saying they "heard voices," thereby increasing public doubts about the
existence of mental ills and worsening the hospitals' already battered image.

Seeing the hospitals assaulted in this fashion, many turned against the
idea that they provided an important refuge. Some decided that the hospitals mostly made patients worse. Some saw drugs, office psychiatry, and
community support services as providing good alternatives. And many, or
even most, came to equate commitment against the patient's will with de-
privation of liberty for insufficient cause.

It made no difference that actively psychotic people with grossly impaired judgment are in no condition to tell whether they need treatment or
a safe haven; or that children and nursing home residents suffer similar
abridgments of liberty for similar reasons; or that military personnel are expected to sacrifice considerable liberty merely because their country says it
needs them; or that people can be thrown in jail for minor antisocial acts.
The fact is that people had lost confidence in the mental hospitals. So in
state after state laws were passed barring involuntary commitment of anyone who was not an "immediate danger" to himself or others, and in case
after case judges interpreted the law so tightly that few people who were
not attempting violence in the courtroom could be committed to a mental
hospital against their will.

What's more, the law followed the patient into the hospital. That is, patients who were not an "immediate danger" to themselves or others could

leave if they desired, regardless of their mental state. And indeed, there was an array of lawyers working for the American Civil Liberties Union and other charitable organizations waiting to help anyone who remained confined when they wanted out. Largely because of these rules, the mental hospitals released most of their patients from 1965 to 1975, and many closed for good.

Of course, by then there were alternatives. Antipsychotic drugs had been around since the mid-1950s; Freudian psychoanalysis was still in vogue; and community support services showed promise. In the minds of many, these things in combination seemed likely to do at least as well as the mental hospitals ever had.

But there were problems. Nobody familiar with severe mental ills and their long-term treatment thought you could simply hand out drug prescriptions and expect patients to get well. Nor did full-bore psychoanalysis seem affordable to most. Instead, the great hope was that community support services, in combination with available medicines and therapy, could do the job. So a good deal of brainstorming was done to see what sort of community support services should be created.

The Massachusetts Mental Health Center played a large role in these brainstorming activities. That was partly because it had impeccable credentials as a community services pioneer and its people ardently believed in the community services concept. In 1920 the director of its outpatient department, Douglas Thom, made the following statement: "Nothing that has happened in modern medicine during the past decade [1910–1919] has been a greater boon to mankind than the movement which transferred medical interest in psychiatry from the insane hospital to the community."[2]

While this claim looks exaggerated in hindsight, there seems no reason to doubt that Thom meant it. And from then on the institution (which in 1956 changed its name from the Boston Psychopathic Hospital to the Massachusetts Mental Health Center) showed that it took the community approach seriously—among other things by developing community care teams, finding ways of reaching the mentally ill homeless, and evolving various day hospital and supervised living arrangements capable of providing long-term care and follow-up for outpatients with severe mental ills. By the mid-1950s it had become the hub of an integrated community mental health system providing a significant part of the Boston area with services that seemed well-suited not only to the city but to the nation as a whole.

Congress gave the community services concept a national stage in 1955 by appointing a joint commission to assess the nation's approach to mental ills. Massachusetts Commissioner of Mental Health Jack Ewalt, who would become Mass Mental's director in 1958, was named to head this study.

The commission's report, issued in 1961, was too subdued for the heady temper of the times. It recommended modifying rather than closing the state mental hospitals. It did press for more emphasis on community and after-care services—including community mental health clinics, halfway houses, social clubs, and sheltered workshops. But it sharply rejected "primary prevention" (prevention of mental ills before they happen by promoting enough social reform to give everyone a healthy environment) as being unworkable.

At the time, this was not what a good many prominent mental health experts and sociologists wanted to hear. It was certainly not what the prevention-touting National Institute of Mental Health wanted to hear. And it was not what President John F. Kennedy, whose sister Rosemary suffered from retardation and mental illness, wanted to hear. So Kennedy appointed a high-level committee to work with NIMH and recommend actions supposedly based on the joint commission report. But the new committee focused mainly on community mental health services and prevention; and the federal government then proceeded to fund a network of community mental health centers (CMHCs) that were not coordinated with the state mental hospitals and not much concerned about the mass of chronically ill mental hospital patients being disgorged.

Furthermore, the share of funds provided to each CMHC by the federal government diminished annually according to a fixed schedule; and as federal funds dried up, the CMHCs tended to limit their relatively high-paid psychiatrists to the one indispensable role of writing prescriptions, replacing them for other purposes with psychologists and social workers. Thus, between 1970 and 1975 the number of psychiatrists in CMHCs fell by half; and instead of spearheading the planned march toward a brighter mental health future, most CMHCs found themselves acting like the fore-runners of today's insurance companies and HMOs, as agents of psychiatry's decline.

At first this had little impact on the "carriage trade" of psychoanalysis, the multitudes of private offices and clinics catering to nonpsychotic people who really wanted psychoanalysis and who could pay. Nor did it bother people who wanted one or another of the new antianxiety medications like Miltown or Valium, because it wasn't hard finding a psychiatrist to prescribe them. Nor did it upset the fundamental mix of services provided to people who were periodically or chronically psychotic, at least not at first, because most of the state mental hospitals were not upended and dumped out suddenly. Rather, their patients were "decanted" gradually. Thus, growth of the severely afflicted but neglected population was also gradual, and for a time it was hard for the press and public to tell whether perhaps severe mental illness might be a myth, or else an ailment created by the mental hospitals. So no alarms went off, and for a long time it was unclear that the only social structure geared to care for severely ill mental patients—the state mental hospital system—was being dismantled without creating any structure able to replace it, and that mentally ill patients who had no parents, relatives, or friends willing to look out for them were being discharged into something like the nightmare conditions preceding the asylums.

As one small consequence of this, the bloom was off the rose at Mass Mental. For like everybody else, the Massachusetts public and Massachusetts state government had lost confidence in the whole idea of state mental hospitals. And while Mass Mental was a lot more than just a state mental hospital, it was seen as one of those. What's more, the Massachusetts economy was doing poorly. So when the state government needed money it took it from the state mental hospital budgets, and of course Mass Mental had the fattest budget. Nor did it help to have a new Massachusetts Mental Health Commissioner, a Californian named William Goldman who was deeply committed to deinstitutionalizing, demedicalizing, and dismembering Mass Mental. In combination, these things marked the end of Mass Mental's rosy love affair with the state and the start of severe pruning. The budget was slashed, the staff was cut, and no new people were appointed to tenured posts.

Jack Ewalt, who remained Mass Mental's director for sixteen years, had trouble adjusting to all this. He really couldn't believe it. He thought the downturn would reverse, and he even thought he could talk the legislature into building a new hospital to replace the original stolid construction dating back to 1912. So he practiced what might euphemistically be called "benign neglect." He didn't fix things when they needed fixing, and eventually the buildings deteriorated to a point where Mass Mental's hospital accreditation was threatened. That put its very treatment of inpatients in question—a big comedown from the days when it was a thriving 300-bed hospital with an eager flood of top-flight residents and a shining reputation as perhaps the leading institution in its field.

Fortunately, as the Mass Mental Health Center was suffering these blows to its previous identity as a powerhouse of psychoanalytic training and inpatient care, it was quietly building an equally strong but more contemporary role as a superior clinical and training facility for community-based care of chronically ill outpatients. It developed a three-tiered model with a psychiatric intensive care unit, a day hospital, and an overnight shelter for patients needing 24-hour structure but not a fully staffed inpatient service; this was complemented by a large outpatient department, the Continuing Care Service, which served the multiple psychosocial and pharmacologic needs of a complicated and socioeconomically bereft population. These services attained a considerable reputation from a clinical standpoint and, later, from a training standpoint as well.

For some years after the decline and fall of the mental hospitals in the 1960s, the Freudians prospered. But eventually the confusion they had helped to sow about the nature of mental ills came back to haunt them. For the benefits of psychoanalysis, however great, were hard to see. So if mental illness could not be defined (making anyone a fit subject for analysis), while the benefits of analysis were hard to see, that made the whole business look both suspect and humorously woolly—as innumerable couch-drawing cartoonists pointed out. And indeed, the psychoanalysts were so far off in their own world that they failed to recognize the growth of brain science that Freud had said would one day replace his theories.

Up to then the theories of psychoanalysis had not been proven or disproven, nor was such proof or disproof likely given the knowledge of the day. But some demonstration of what psychoanalysis could do was badly needed, not only because psychoanalysis was prolonged and costly, but also because the psychoanalysts had shown themselves to be inflexible in the face of criticism. Instead of welcoming the rush of new ideas coming in from brain science and elsewhere, all too often they had pulled their wagons into a circle. As Edward Shorter notes in *A History of Psychiatry*: "The demise of psychoanalysis was in large measure a result of its own lack of flexibility, its resistance to incorporating new findings from the neurosciences. And this reluctance was directly related to the analysts' fear of being proven wrong."[3]

This fear had been growing for some decades—at least since the 1940s. But not everybody was afraid; indeed, many psychoanalysts were eager to confront public skepticism by showing what psychoanalysis could do. So in 1948, and again in 1953, the American Psychoanalytic Association set up committees to evaluate psychoanalytic therapy. While the first ended its work after running into "insuperable resistance among the membership," the second was more persistent. It actually gathered 3000 reports from members as well as follow-up reports on the analytic outcomes of these cases. However, at this point great difficulties were encountered. Most of the collected data were lost or vanished; the committee was disbanded; and all but a few minor threads of the evaluative effort ended.

Nor did psychoanalysis fare well in outside studies. One early work, published in 1952, compared the results of psychoanalysis and other "eclectic" therapies on neurotic patients. The findings indicated that 64 percent of those receiving the eclectic therapies had improved—a figure comparable to the percentage expected to improve without treatment—whereas only 44 percent of the psychoanalytic patients were found to have shown improvement. In other words, the psychoanalysis appeared, if anything, to have hindered progress.

Not surprisingly in view of such results, as the years passed psychoanalysis found itself facing progressively tougher competition from other psychotherapies. For the truth is that public confidence in psychoanalysis was waning at a time when public demand for psychotherapy was growing and public information about the nature of mental ills was confused. As a result, psychoanalysis saw many clients and potential clients wander off into a jungle of competing psychotherapies, some with no academic standing whatsoever and some involving pretty bizarre things like reincarnation, invocation of "natural" forces, or "therapeutic touching."

Of course, the retreat of psychoanalysis also gave encouragement to other therapies that seemed to make more sense and that were considerably more subdued. That may have come as a relief to many seeking sound and modest therapy. However, it did little to help psychiatry find its way in the changing mental health world, because most practitioners of the new methods were counselors, therapists, or psychologists rather than psychiatrists, and most had few or no formal ties to either biomedicine or brain science.

One of the most powerful influences driving these trends was the fact that psychoanalysis was both time-consuming and expensive. It was especially expensive because virtually all analysts were psychiatrists, and psychiatrists earned pay commensurate with their long and expensive medical education. Indeed, most people could not afford psychoanalysis unless they were rich or unless the government or their insurance companies paid the bills. So, as federal funding for the community mental health centers fell off, and as private insurance companies became increasingly unwilling to pay for extended analysis sessions, the world of psychoanalysis shrank more and more, until it became a small fraction of its former size.

On another front, psychoanalysis was facing competition from antianxiety drugs that appealed powerfully to the general public. "Tranquilizers," led by Miltown, burst on the scene in 1956, and soon one American in twenty was taking them. The stronger benzodiazepines—Librium, Valium, and others—followed in the 1960s. By 1970, long before the even more popular Prozac arrived in 1987, one woman in five and one man in thirteen was taking some kind of antianxiety medication.

The implications of all this for psychiatry were considerable, because unlike the powerful neuroleptics useful against psychosis, these drugs helped to deal with far commoner anxiety-related problems. And while the benzodiazepines proved quite addictive, they also proved their worth; so they remained in common use and accounted largely for the fact that by 1975 a quarter of all office visits to psychiatrists produced a drug prescription.

Meanwhile, despite heavy losses elsewhere, psychoanalysis kept proclaiming its worth against the major psychoses, most notably schizophrenia. But in the latter 1960s there came a dawning realization that psychiatric drugs could deal with severe mental problems better than psychoanalysis, and that in fact treating psychotic patients with psychoanalysis alone was all but useless.

This idea did not arise in the mid–1950s with the first antipsychotic drugs, because everybody knew the drugs just treated symptoms without curing anybody, whereas the psychoanalysts were claiming to cure people. For instance, a chief resident at Mass Mental commonly referred to the antipsychotic Thorazine as a "chemical straitjacket" because it allegedly crippled the patient's ability to make a curative transference. He was just one of innumerable analytically oriented doctors who discouraged drug use because they felt it interfered with their own treatments. And so, for a while, the drugs were seen as a second-rate fill-in for the best type of treatment, which remained psychoanalysis.

One of the investigations that changed this view, published in the *International Journal of Psychiatry* in 1967, was led by Lester Grinspoon, then a staff member at the Massachusetts Mental Health Center, and Jack Ewalt. Both of these men then believed analysis to be effective against schizophrenia, but they were bothered by the failure of Mass Mental's residents to produce positive results. Their theory was that the residents simply lacked experience. So they got an NIMH grant to hire some top-flight Boston analysts and do a study that would compare the results of psychoanalyzing chronic schizophrenics and treating them with the antipsychotic thioridazine (Mellaril). As Grinspoon explained in a recent conversation,

This was the first study to look into the question of psychotherapy versus pharmacotherapy. The curious thing about the study is that ethically you couldn't do it today. Because what we did was take a group of patients with chronic schizophrenia and randomly distribute them so as to test the effects of thioridazine and psychoanalytically oriented psychotherapy. So I randomly assigned the patients to a group that got the thioridazine or one that didn't, and there was no problem with that. But when I tried to do the same thing with the therapy variable, using some people at the Harvard Divinity School to serve as sham psychotherapists, the powers that be of the hospital said "No, no, that would be unethical. Psychotherapy is what's going to help them overcome their illness. You can deprive them of drugs, because drugs don't really cure anything, but you can't deprive them of psychotherapy." This was just the reverse of what the attitude would be today.

Because of this we had to give all the patients psychotherapy, and the work was less than ideal for that reason. Even so, it served to show that the drug was beneficial while the effects of therapy were nil. That is, the patients receiving thioridazine clearly did better than those who went without it; and at the same time none of the data indicated directly that psychotherapy was helpful while a good deal of it suggested indirectly that psychotherapy made no difference. For example, one of the participating psychoanalysts died, but his patients went on just as before. We also had loads of data such as diary material, and there was no way anyone could tease out of those data findings suggesting any influence for psychotherapy whatsoever.

These results were all the more dramatic because the participants were some of Boston's leading analysts. So together with another project done at California's Camarillo State Hospital and reported in 1969, the Grinspoon–Ewalt study opened the floodgates. Other studies followed, and soon the claim that psychoanalysis could cure severe mental ills lay in ruins. Grinspoon himself later abandoned psychoanalysis and resigned from the Boston Psychoanalytic Institute.

Over time such developments, combined with the analysts' continuing failure to confirm benefits, cut the ground out from under psychoanalysis.

In the academic world of psychiatry, psychoanalysis was dethroned. By the mid-1970s analysts were only rarely being selected to head psychiatric departments or hold professorships. The supply of psychiatrists interested in becoming analysts dwindled; and the long hours of training that psychiatric residents had once received in psychoanalytic psychotherapy were cut drastically or abolished altogether.

Of course, the psychoanalytically trained psychiatrists did not simply go into a big closet somewhere, to be replaced by drug-oriented psychiatrists and nonanalytic therapists. Rather, most of the analysts and analytically oriented psychiatrists of the day got interested in drugs and alternative therapies. Indeed, a 1974 poll by the American Psychoanalytic Association revealed that three-fifths of the analysts surveyed had begun prescribing medications and a third were offering some kind of nonanalytic therapy.

That's just as well, because soon the retreat of psychoanalysis turned into a rout. Droves of scholars entered the lists against the Freudians. Nearly all of Freud's central concepts—his theory of dreams, the Oedipal complex, infantile sexuality, penis envy, childhood repression, and so on—were cast aside. Never having been proven, most were not really subject to disproof. But they were no longer taken seriously by most psychiatrists, and so they were placed in the same sort of scientific limbo as astrology.

This slow decline of psychoanalysis exposed and inflamed psychiatry's split personality. For the cloak of invulnerability and mystery surrounding psychoanalysis had been lifted, leaving two courses to pursue. These two courses were biomedical treatment (mostly medical diagnosis and drug prescription) and "humanistic" therapy (treatment based on interactions with the patient, plus discussions with the patient required for understanding the patient's case, establishing a good and caring relationship, guiding patient activities, monitoring progress, and coordinating social services and follow-up). In the case of almost any severe mental illness, both of these things are needed for sound treatment. Moreover, as we have seen, the two are inseparable—because sound biomedicine demands a good understanding of the patient's case, periodic monitoring, and a retailoring of drug treatments to changed conditions, while sound therapy requires a thorough understanding of what biomedicine can do, and also of possible problems it may pose.

However, from the viewpoint of the community mental health centers and also the insurance companies and HMOs that came to play an increasing role in paying for treatment, only biomedical treatment was deemed cost-effective. Therapy, especially the intense, interminable therapy favored by psychoanalysts, seemed too expensive. So the high-paid psychiatrists were limited more and more to medical diagnosis and drug prescription, while therapy was assigned to social workers, counselors, and psychologists who were cheaper and less committed to time-consuming methods.

The "split personality" disorder thus induced within the mental health field did not devastate most psychiatrists who were already established. Their reputations were intact, their services in demand, and their options open. So, without concerning themselves overly about theoretical integration, they could decide whether to embrace biomedicine, continue with psychoanalysis, begin using other kinds of therapy, or pursue two or even all three of these alternatives.

But the disorder struck hard at new psychiatrists. They had no established reputation or client base. So they found themselves penned in. They could not do therapy and hope to pay for their medical education. Instead they had to depend on the CMHCs, HMOs, and insurance companies for compensation. And that meant they had to limit themselves to short appointments, commonly as short as 20 minutes, dedicated to prescribing drugs.

Because this is neither a pleasant nor a sound way to practice psychiatry, the impact on recruitment was catastrophic. Whereas in the 1950s and 1960s psychiatry had attracted the best U.S. medical school graduates to its banner, by the 1980s enthusiasm had dimmed sharply, and by 2000 it had practically vanished.

This point came up in a recent conversation with Dr. Judith Rapoport, chief of the child psychiatry branch at the National Institute of Mental Health. As she put it: "I wouldn't recommend to anybody that they get a residency in psychiatry today, because jobs with HMOs require writing three prescriptions an hour for people who, if they wish to talk to someone, will wind up talking to a therapist paid at about half their going rate."

More specifically, she observed that every year "there are about 1200 first-year residency positions [for psychiatrists] available nationally. For the last four years, about 375 of the applicants have been American medical school graduates, most of them in the bottom quarter of their medical school class. So the match is made on a Monday in March, and about 95% of these people get taken. By Wednesday the pool of foreign medical school graduates comes in, and so the remaining 70 percent of the positions have been filled by foreign medical graduates. Generally speaking, these are people from developing countries. Their pass rates at the [required] board examinations are low. If this does not change, psychiatry could disappear as a medical specialty or could be merged with another specialty such as neurology."

The overall effect of this trend, besides reducing the quality of drug treatment, has been to widen psychiatry's split personality. For psychiatrists are increasingly confined to the job of writing prescriptions, while psychologists, having no medical background, are limited to "humanistic" therapy. As Dr. Rapoport noted: "If you were an HMO you'd be foolish to hire them [psychiatrists] to do anything else [but write prescriptions] when 70 percent of the residents in psychiatry are foreign medical school graduates who may not be fluent in English and who have a different culture. Whereas the psychology and clinical psych programs have about 20 applicants for each position. These are excellent students from all the best schools in the United States. It's a far better pool to get therapists from, and they're half the price."

Probably because psychiatry has sunk so low, it is now facing serious competition on its home turf. Not only are psychologists periodically clamoring for the right to prescribe psychiatric drugs, but general practitioners have become leading prescribers of Prozac and other antidepressant medications.

The main point, however, is not that competition exists but that therapy has been divorced from medicine. As things stand now, most therapists are limited to an anecdotal knowledge of psychiatric drugs and anecdotal awareness of what we have learned about the brain; while our new pre-scription-writing psychiatrists have an uncertain command of medicine and often know far less than they should about their patients. Thus, very much like a victim of psychosis, psychiatry has lost much of its ability to re-late to the real world and do its job. The result is understandable: Anyone seeking mental health care today runs a high risk of getting low-grade treat-ment. Beyond that, millions of people with severe mental ills are being im-properly treated or neglected; coordination and follow-up efforts on their behalf range from poor to nil; and degeneration within the field has reached a point where we now have a choice of actively rebuilding psychiatry or else being prepared to see it fall apart.

This last option has serious drawbacks, because no other specialty could fill the gap left by psychiatry's demise. The nearest fields are psychology and neurology. But psychology lacks psychiatry's medical and hard science base, while neurology lacks command of any strong mind therapy sensitive to human needs. We will have more to say about this later. But the bottom line is that neither psychology nor neurology seems prepared to effectively take over the ground lost by psychiatry. So the logical result of further dis-integration would be more neglect of the mentally ill—and we already have too much of that.

That means it is high time for psychiatry to be rebuilt. But the building must be done on a new foundation, because the psychiatry of the asylums and the psychiatry of Freud were built on terrible foundations. Asylum psychiatry was built on a desire to care for the mentally ill without knowing how to do it; and Freudian psychiatry was built on guesswork. All this was understandable, given the circumstances of the times; but it is a far cry from where we are right now.

For we have in fact reached solid ground. Our brain science base is strong. As a result, we are in a position to replace psychiatry's wobbly foun-dation with a firm one, develop a new psychology compatible with brain science, heal the profession's split personality, and for the first time create a sound well-supported structure. Our own conviction, one that we believe will be upheld by any careful and objective review of the situation, is that the rebuilding of psychiatry needs to be founded on a clear public and professional understanding of certain basic points: that drugs by themselves are not enough; that tailoring drugs to particular cases is not simple; and that the psychiatrist needs to see the patient long enough to understand the case, form a sound therapeutic relationship, and pave the way for effective interaction with the close-knit framework of workers required to provide hospital care and therapy, guide living arrangements, monitor conditions, and make sure the patient is not lost to follow-up. Of course, all this can be expected to cost money. But it will be considerably cheaper and yield far better care than the ill-coordinated mess of medical, psychologic, therapeutic, counseling, HMO, legal, police, jail, hospital, and homeless shelter costs that we are paying right now.

Another basic point is that the main support for both the drug and therapy sides of this rebuilding effort is provided by brain science. For we really are beginning to learn how the brain works, not just how neurons work but how things like sight, hearing, language, thought, memory, and emotions work; how the brain regulates its conscious and unconscious states; what brain substructures like the cerebellum, brain stem, hippocampus, amygdala, and thalamus do; how these and other substructures interact; how our psychiatric drugs work; and how theories about various chemical, structural, and coordination problems within the brain can help to explain a wide range of mental ills.

In most branches of medicine we assume that doctors know how the relevant organs and systems of the body work, and that this contributes to their skill in treating patients. Brain science is now reaching a point where it can provide such knowledge to psychiatry. Clearly, it does not have all the answers, and given the brain's complexity it presumably never will. But it has enough answers to deeply enrich our understanding of mental ills and provide an important guide to treatment.

That makes it important for psychiatrists to have a sound understanding of brain science. What especially encourages movement in this direction, even among older, psychoanalytically inclined psychiatrists, is the fact that brain science has matured and so is not as hard to grasp as it once was.

Typically, new scientific disciplines are hardest to learn when they are young. At first little is known, most results are either hedged or confused, few teaching routines have been established, and those who would learn must be prepared to leap from one small island of information to another. In time, however, a good-sized body of information emerges, basic concepts evolve, and learning becomes easier. This doesn't mean one can get a professional grounding in brain science by reading books like this one, or like Francis Crick's *The Astonishing Hypothesis: The Scientific Search for the Soul*, Joseph LeDoux's *The Emotional Brain*, Allan Hobson's *Dreaming as Delirium*, or any other popular or semipopular book about the brain. It simply means that the subject has been getting easier, and that many topics that seemed daunting for all but a few scholars and researchers twenty years ago are now well within the reach of general readers.

All this has both advantages and drawbacks. A prime advantage is that we can use this knowledge to rebuild psychiatry. A prime drawback, from the standpoint of some psychologists, is that we must do so. We have no choice. For the progress of brain science has drawn the attention of our scholars, press, and public like a strong magnet draws iron filings. No top-down humanistic psychology can hope to compete with that. So we cannot expect to replace Freud's century-old theories with something else unless that "something else" is grounded in brain science.

We think we know how psychology might be successfully blended with brain science, and also how psychiatry could be rebuilt. We shall present these ideas in due course. But we also think it an error to invoke the blessings of science in name but not in substance, and we have no wish to exchange tirades with anti-science devotees about the alleged progress or nonprogress of brain science. So we think it would be useful to take a look at some of the things that brain science has discovered. Of course, any good scholar could fill a huge library with the available literature; but our space and time are limited, so our tour will necessarily be brief. We will deal only with certain subjects—consciousness, sleep and dreams, drugs, anxiety disorders, depression, and schizophrenia—and will give only partial coverage to these. Even so, we think this short tour will show that the field of brain science is ready—that it can guide our efforts to restore psychiatry, promote a balanced blending of mind therapy with medicine, and provide the sound foundation of knowledge that we need.

We realize that until now we have been talking mostly about historical and social issues, and that our shift of focus to brain science may therefore seem abrupt. But it is vital to get a real as opposed to theoretical look at what brain science has been doing. So even though we expect to make our journey into the brain and mind enjoyable, it is not an incidental adventure. Rather, it is at least as important to understanding our case for reforming psychiatry as Herman Melville's discourse on the whaling industry was to the development of *Moby Dick*.

PART TWO

Finding the
Mind's Brain

5

Consciousness:
The Master Magician and
the Waterfall

*The scientific belief is that our minds—the behavior of our
brains—can be explained by the interaction of nerve cells
(and other cells) and the molecules associated with them. This
is to most people a really surprising concept. It does not come
easily to believe that I am the detailed behavior of a set of
nerve cells, however many there may be and however intricate
their interactions. Try for a moment to imagine this point of
view. ("Whatever he says, Mabel, I know I'm in there
somewhere, looking out on the world.")*

—**Francis Crick**,
The Astonishing Hypothesis, **1995**

In the early 1960s, the mathematics fraternity in this country became concerned about the quality of math teaching in the nation's schools. The Cold
War needed engineers, the Russians had launched *Sputnik*, and high school
math instruction was felt to be subpar. So leading mathematicians devised
something called the "New Math" and leading educators sold it to the schools.

The result was a disaster. The New Math introduced a host of new concepts—about sets, numbers theory, and so forth—that interested mainly

mathematicians. It had little relevance for students, many of whom would use math mainly to add grocery lists and balance checkbooks. It was hard to teach, often hard to understand, and it created such confusion that instead of improving math teaching, it set back both math instruction and the average student's ability to do math quite a lot.

Recalling this experience, and noting that we, your authors, delight in brain science, it seems worth asking whether we have let our natural enthusiasms run away with us, or whether in fact psychiatry needs a far stronger brain science foundation than it has now.

We believe that it needs a stronger brain science base for several reasons. First, this inclination to stress brain science is not arbitrary. Unlike the New Math advocates, who embraced knowledge that had been around for centuries, we are pressing for use of information that is new—most of it developed within the last three decades. The vast bulk of this knowledge was not available either to the asylum psychiatrists or to the Freudians, which helps to explain why both of them ultimately failed.

Second, psychiatry's brain science base is very weak. Most college and medical school students who home in on brain science today go into neurology. And most of the more advanced training for psychiatric residents is really apprenticeship training in which brain science plays little or no part—even when it comes to biomedical training in how to diagnose mental ills and prescribe drugs. So the brain science knowledge of many practicing psychiatrists remains mostly informal or even anecdotal, leaving psychotherapy and psychopharmacology separated, isolated, and diminished at a time when brain science has the ability to nourish and combine them in an empowering fashion.

Third, brain science is vital. It is telling us what the psychologists—including the Freudians—have always dreamed of knowing. It is telling us how the mind works. So it has all kinds of applications. Some of these applications are quite specific. As the case of Alfred Ramsey will show in Chapter 11, understanding the brain science basis for specific drug actions can help us tell whether one psychiatric drug might be a fitting replacement for another in treating a schizophrenic or whether it would likely do more harm than good. Brain science knowledge can also indicate why a given psychotherapy might work for a particular patient but not others, and can explain how dream reports might help us to understand a patient's plight. More broadly, brain science can show why psychotherapy and psychopharmacology need to be intertwined, why mental disorders tend to be chronic but also have dramatic ups and downs, how sleep and dream patterns relate to mental ills, what causes various kinds of anxiety and depression, why mental ills as diverse as schizophrenia, depression, mania, and various anxiety disorders sometimes coexist in the same patient, and even why the disorders afflicting particular patients sometimes drift around from one diagnostic category to another.

We could go on indefinitely. The list is endless. For in fact brain science is providing myriad explanations for the mind's work in concrete terms. This compelling platform of knowledge, for lack of which Sigmund Freud's psychoanalysis crumbled, is the foundation upon which modern psychiatry should stand.

Let's look at this another way. Traditionally, psychologists and psychotherapists gain access to a problem through the patient's conscious mind, that is, by talking with the patient. Various techniques are then used to assess unconscious processes; but since such assessment is indirect, these unconscious processes tend to remain mysterious. A prime job of any psychology, including Freudian psychology, is to get around or dispel this mystery of what the unconscious mind is doing.

For a brain scientist the reverse is true. As shall be seen, conscious processes within the brain are limited, accounting for only a small part of the brain's work; and brain researchers typically have neither the wherewithal nor the motive to separate conscious from unconscious processes. Hence, they rarely make this distinction and tend to consider consciousness a mystery.

This inverted state of affairs might appear to make brain science incompatible with both psychology and the clinical practice of psychotherapy. But in fact the opposite is true, because brain science has just what the psychologists and psychotherapists want—good access to the unconscious mind.

Of course, as we have seen, the unconscious data-handling processes charted by brain scientists often look quite different from those once envisaged by theoretical psychologists. But since the unconscious mind of our myriad theoretical psychologies was always wreathed in mystery anyway, major differences between theory and reality were to be expected; and since brain science has now matured to a point where it can and should play a vital role in clinical practice, there is good reason to develop a neurodynamic psychology based on brain science that can help balance biomedicine with therapy and encourage sound treatment.

Putting it more urgently, we need to strengthen psychiatry's brain science base because, as we have seen, psychiatry doesn't need a minor tuneup. It needs resuscitation. Psychiatry is dying. Its stream of able and enthusiastic recruits has dried up, and multitudes of its most severely ill patients are being neglected. So we must act. Doing nothing is no option. But even if we wanted to, we could not reinvent psychoanalysis or find some other speculative psychology to unify and revive psychiatry, because nothing of that sort can hope to rival the bright light now being shed upon the mind and mental ills by brain science. Instead, we should harness the source of that light to our purpose.

To this end, we are about to examine some of the things that brain science can tell us about mental ills. So how should we begin? Should we simply dive into various disordered states, or would it be better to see how the normal brain is organized—how it processes the sensations, thoughts, and feelings we experience every day with our conscious minds—before proceeding on to mental ills?

We have chosen the latter course for two reasons. First, it allows us to present certain basic points about how the brain works that will be used later. And second, it gives us an opportunity to probe the age-old "mind–body" problem of how the brain generates the mind—and so to explore the nature of human consciousness—from the standpoint of brain science.

This has a direct bearing on mental ills, not only because all mental ills affect the conscious mind, but also because consciousness is the main window used by all psychologists, therapists, and counselors to view and treat the mind. We realize that Freud, Jung, and others paid a lot of attention to supposed preconscious and unconscious processes around the dawn of the last century. But they only gained access to these theoretical processes by talking to their patients. So, like their modern counterparts, virtually everything they did involved consciousness. And thus the study of the mind that is psychology—whether it be psychoanalytic, behavioral, cognitive, or some other theoretical variant—has always been pretty much content to access the mind through the window that consciousness provides, and to spin out humanistic theories that are not directly beholden to brain science.

We think this approach is weak. We think the time has come to include brain science directly. For that reason we are proposing the development of "neurodynamics," a new psychology grounded in brain science that will be oriented to the assessment and treatment of mental ills. We will have a lot more to say about this later. For the moment, however, we need to be very clear about how the brain and mind relate to one another. For that reason we have chosen to start by describing some of what brain science has to say about human consciousness.

We recognize, of course, that throughout history the nature of consciousness has been an enormous riddle. Even today, we can't really put our finger on what consciousness is. And while we all experience it, we can't see or touch it, and we don't know precisely how it works. It is as though we were in the thrall of a master magician, destined to be lifelong members of a fascinated audience utterly baffled by his commanding sleight-of-hand.

What's more, many brain scientists have tended to avoid the subject. That's partly because it is often hard to distinguish between conscious and unconscious activities in the brain, and partly because nonscientists have dealt with consciousness so long that the subject looks unscientific. Nevertheless, things are not so obscure as they might seem. Indeed, those brain scientists who have chosen to get involved are fast accumulating a critical mass of direct and indirect information about consciousness that seems capable of unraveling its mysteries.

Before turning to that, however, we should first seek to define exactly what we mean by consciousness. That's not easy. Though we are all conscious most of the time, other creatures are conscious too, and it's hard to tell how much of our sort of consciousness they share. We are also conscious in different ways at different times, as when we are having a nightmare, solving a math problem, or hitting a home run. Moreover, the word is used in everyday language to deal with all sorts of side issues like consciousness-raising and Krishna consciousness; no generally accepted scientific definition has emerged; different authors use different meanings or skirt the issue altogether; and so anyone thinking that some simple definition will suffice should beware.

In essence, consciousness is awareness. But plants can be "aware" of the sun, as they demonstrate by following it around, though few except whimsical romantics would ascribe any real consciousness to them. To match the common understanding of consciousness, awareness must be elevated at least to a point where the conscious creature is aware of itself, as distinguished from the rest of the world; is also aware of the outside world to some degree; and can use perceptions of that world to act on its own behalf.

All of this requires an information processing system—in other words, a brain. According to this definition, receiving information raw from nature does not produce consciousness. Instead, the information must be processed until it becomes coherent and acquires meaning. This implies that consciousness is not any old awareness. Rather, it is awareness of information processed by the brain.

There seems no reason why brain science cannot explore this concept as it applies to ourselves. But anyone who would explore human consciousness must deal with more than just brain science. For this ancient riddle has bedeviled generations of scholars back at least to the dawn of human history, and so we must address not just scientific questions but also religious and philosophical views that still color our thoughts about consciousness today.

One such view is that the conscious mind may somehow reside outside the brain altogether and may owe its existence to forces of which we are unaware. The best-known advocate of this theory, called "dualism," was the seventeenth-century philosopher René Descartes, who suggested that the conscious mind communicated with the brain through a small organ near the brain's center called the pineal body. Eventually people learned that the pineal body was more humble—among other things secreting melatonin and inhibiting human gonad development. No evidence appeared supporting Descartes' view, and so in due course that view was rejected.

Nevertheless, dualism shorn of the pineal body lives on. Quite aside from any theoretical merits, it has proven an attractive idea for many religious thinkers, because the immaterial conscious mind provides a ready avenue to immortality after the brain dies. Dualism also appeals to a number of quantum physicists, who see dualistic mental dependence on quantum effects as a good vehicle for resolving certain mysteries of the subatomic realm. Though it looks odd seeing religious philosophy and quantum physics hitched to the same wagon, scholars don't care much about that. But they do care about a big problem that makes dualism hard to defend. The big problem is this: the mind must be so immaterial as to evade detection. However, controlling the brain means pushing things (electric charges and brain chemicals) around. So how can the mind retain its immaterial (undetectable and nonphysical) nature while doing all this detectable physical work at the same time? This difficulty, which has been recognized at least since Descartes' time, is widely regarded as dualism's fatal flaw.

Partly for this reason, most people who believe mental processes involve something more than just the brain and nervous system take another tack. Some, like noted mathematician and physicist Roger Penrose, claim that a better grasp of quantum physics is needed before full understanding of the mind can be gained. Others, like author John Horgan and philosophers Colin McGinn and Thomas Nagel, suggest that the mysteries of the conscious mind surpass anything we can ever grasp. In general, such people highlight the conscious mind's mysterious and unknown sides while stressing the obvious self-analysis problem—much like the humorist who once said, "Of course the conscious mind is complex! If it were simple we couldn't understand it!"

What lends strength to these obscurantist views is the fact that we are trapped inside our own heads. It is hard to accept the idea that we deal only with nerve impulses. Instead, we are so attuned to this established process that we accept those impulses as reflecting reality without noting that much of the information received is highly processed. When you read this sentence, for instance, you focus on what the sentence means. You do not attend to the shape of individual letters, the process of grouping letters into words, or the mental gymnastics used to tie words and clusters of words to specific meanings.

This unfounded assumption that we are in contact with raw reality makes it hard to grasp that what registers in consciousness is not raw reality, but instead is limited information received, selected, translated into abstract neural codes, and processed by the brain. It is of course true that we currently lack the scientific knowledge needed for complete understanding of human consciousness. But here we are dealing only with the basic concept—something that should be quite easy to apply to ourselves except for the fact that we are not set up to do it. Taking a phrase from Walt Kelley's midcentury cartoon character Pogo, when it comes to effectively comprehending the nature of our own consciousness, "We have met the enemy and they are us."

A good way to start clearing the air and dealing with this problem is to distinguish what we mean by "consciousness" from what we mean by "mind." Basically, the difference is that "consciousness" requires awareness and so is rather limited, whereas the generally accepted concept of "mind" includes all of the brain's mental activities, whether they are conscious or not.

We can see how consciousness is limited if we look at the role it plays in some chosen action. For instance, consider what happens when a tennis player hits a serve. The server starts by tossing a tennis ball ten feet or so into the air, eyeballing it, and bringing the racket around behind his or her back. Then, without ever looking at the target area, the player strikes the ball, sending it twenty feet over a three-foot net and into a small service court, hoping it has enough speed and spin to thwart the opposition.

Is this act wholly conscious? It seems conscious, and the decision to serve the ball is surely conscious. But as anyone knows who has tried it, serving takes practice. In fact, it takes a chorus of minor skills—like knowing how to move your feet, hold the racket, toss the ball, and bring the racket behind you in a tension-building arc. You must also know how to turn this built-up tension into racket speed as you bring the racket up, and how to hit the ball precisely enough to make it sail over the net and land in the right court.

If you play tennis often, you should be able to perform each of these small acts consciously. What you cannot do is call up each one consciously and run through them all consciously in the two or three seconds between the time the ball leaves your hand and the racket strikes. So clearly, you are doing a lot more than can be accomplished consciously in the time allowed.

What seems to be happening is this: your brain learns each of the needed acts by trial and error. The results get stored as memory trains (closely linked memories), of which you remain oblivious unless you choose to examine them consciously. Good ways to coordinate these acts also get learned and stored. Then, when you decide to serve, you summon a well-coordinated sequence of memory trains to do the job. This suggests that about the only conscious things you do in serving are to call up a specific series of these memory trains and then look out for variations (slippery court surface, ball in the wrong place) that demand adjustment. Furthermore, since your chances of successfully adjusting something badly out of whack in a split second are not good, your conscious mind is a good deal less than the all-powerful manager it seems. Instead it is mostly an onlooker, a more potent onlooker than Queen Elizabeth at Wimbledon, but an onlooker nonetheless.

Or else, remember how hard it was to shift gears when you first started driving a stick shift? Today (if you are a licensed manual shift driver) you find it easy, so easy you don't have to think about it consciously, because learned memory trains and associations do the work. Indeed, there is good reason to think that your unconscious mind, directing these processes, does most of your driving, often leaving just enough conscious awareness in place to detect variations (trucks, taxicabs, red lights) that need attention. That's why you have plenty of conscious mental capacity in reserve to chat with riders.

But matters don't end there, because chatting is like driving. It depends on memory. It relies on learned memory trains and established data processing channels—in this case very complex ones—to divine the meaning of incoming syllables, words, and sentences, and also to turn your conscious responses into speech. This is true not just of chatting but of all acts involving language. Beyond that, we can see that virtually all other kinds of learned behavior use memory, and so memory is deeply involved in just about everything one would normally consider "conscious" acts.

What we find, then, is that while the line separating conscious and nonconscious parts of "conscious" acts is hard to draw, consciousness is indeed limited. Even so, the arrangement looks reasonable. We can understand why it should be so, and can easily imagine the linking of consciousness with various other operations in the brain.

In this vein, we should note that just as consciousness is not the same as memory, it is also not the same as thought. That is, consciousness is awareness, and awareness need not involve thought. If you are driving and come to a 🛑 sign, the shape and color of the sign, its letters, and the sign's meaning would all flood into your consciousness quickly, following established data processing channels involving little or no thought. On the other hand, if the sign's color and shape were the same but the letters read 🛑 you would need a moment to think before realizing "Oh! The sign is upside down!" In the first case, what we generally regard as thought would be minimal or absent. So clearly, although consciousness can involve thought, as in the case of the inverted sign, it doesn't have to.

This point deserves attention, because it departs from the current view of many experts. Perhaps that's because throughout much of history consciousness was equated with the soul; and since it's hard to imagine a human soul with no ability to think, this concept tied consciousness to thought. We may simply have hung onto this old idea. Or we may be trying to define our special nature. That is, we are all proud of our human status and like to distinguish ourselves from other creatures; so it seems attractive to imagine that we are the only "conscious" ones around. That purpose is served if we can say that consciousness requires our kind of thinking, because then consciousness is limited to us.

But these things are just desires that do not necessarily reflect the truth. What's more, if we bind consciousness to thought we find that it creates a lot of trouble. For as we have seen, we appear to be conscious of lots of things involving little or no thought. Furthermore, human infants cannot use the frontal lobes of the cerebral cortex, where most thought processes occur, because these lobes are immature. Does this mean that babies are not conscious? We can even take this one step further, because the prefrontal lobes that serve as centers for our most involved thoughts do not mature until late into the teenage years. And despite occasional contrary claims by pained parents, the idea that children and young teenagers are less than fully conscious clearly exceeds the bounds of common sense.

If we go back to our definition of consciousness as awareness of information processed by the brain, we can see why thought isn't needed. For your brain processes lots of things—vision, hearing, language, thought, emotion, and so forth—of which thought is only one. You can be aware of any of these items, but you need not be aware of any particular one to be conscious. Blind people and deaf people are conscious, and so are people when they are startled by lightning flashes, consumed with rage, or serving tennis balls—not just when they are engaged in higher thought.

Of course, accepting this definition implies that virtually all other creatures are conscious too. That's because nearly all have some sort of brain, and nearly all act as well-coordinated individuals. So nearly all must have some brain system that registers awareness of what the brain is finding and manages a unified response. That means humans, apes, and dolphins are not alone. Dogs and cats are conscious; so are mice; and so are snakes and rats. Indeed, according to our definition it seems likely that even cockroaches are conscious.

That doesn't necessarily increase our regard for any of these creatures, because the nature of any creature's consciousness depends on the nature of its brain. For example, one cannot imagine a consciousness employing language as we know it that does not possess the ability to make internal speech. More generally, our own consciousness, which among other things includes a finely honed awareness of that consciousness itself, as well as of personal identity, time, language, vision, and abstract concepts, depends on our brain's well-organized horde of accumulated data. Were that horde not so huge and handy—witness the plight of everything from chimpanzees to cockroaches—our sense of consciousness would dim, and our potential range of conscious acts would shrink.

This suggests that the human brain's power is worth assessing. But making such an assessment, even a very rough one, isn't easy. We didn't build the brain, so we don't have its engineering plans. Even so, we can see that anything managing the intricacies of speech, vision, memory, abstract reasoning, and many other impressive functions without obvious strain must be powerful. And that provides an incentive to find some sort of primitive yardstick for comparison.

One such primitive yardstick that humanity *has* built is the computer. Unfortunately, as will be seen, the brain works very differently from a computer, and we have no accurate way of comparing their operating abilities. Nevertheless, to get a very rough idea of what is happening, one might reasonably compare their data-holding capacities, keeping in mind that what we know of the brain's data-holding capacity is limited, and so the results of any such comparison will necessarily be crude.

Suppose that for purposes of this comparison we select a year 2001 desktop computer with 250 megabytes of RAM (random access memory). A byte is equal to 8 "bits" of yes or no information (on-off switches), so in theory the computer's memory might contain some 2 billion (2,000,000,000) bits of yes or no information.

In the human brain, the nearest equivalent to the computer's on-off switches are the small gaps called synapses between the neurons (nerve cells) that receive and transmit signals. We aren't sure how many neurons an ordinary human brain contains, but most experts seem to favor an estimate of roughly 100 billion. In addition, the average brain neuron has something on the order of 10,000 synaptic connections with other neurons, and other things like variations in the strength of signals across synapses, different types of neurotransmitters and receptors, and variable neuron firing rates could well increase the brain's data-holding capacity 100 times or more. Multiplying all these things together suggests that the average human brain's data-holding capacity is something on the order of 100 quadrillion (100,000,000,000,000,000) bits of information.

Dividing this crude estimate of the brain's capacity by that of our computer suggests that the human brain's capacity exceeds the computer's by a factor of something like 50 million. That is a huge difference. Suppose, for instance, that our desktop computer is in a case eighteen inches long. In that event, 50 million of them placed end to end would form a line of computers stretching over 16 *thousand* miles—two-thirds of the way around the planet. In other words, the human brain's massive computing capacity, evolved by Nature in something the size of a grapefruit, makes the best work of our microminiaturizing computer wizards look like the first uncertain steps of a small child.

This vast computing power helps to explain a lot of things. For one thing, it explains how consciousness can be so limited—reflecting only part of the brain's work—and still be so full of richly detailed sights, sounds, feelings, memories, thoughts, and so on.

Of course, even the human brain can only do so much, which probably explains why it receives only limited sensory inputs. Our sense of smell is poor. We don't hear all sounds, which is why dogs can hear high-pitched noises and faint noises that simply don't register with us. We don't sense the polarization of light like bees do, or spot small rodents from thousands of feet up like eagles. Nor do we sample more than one area, the small "visible light" portion, of the electromagnetic spectrum, which is why X-ray, ultraviolet, infrared, microwave, and radio radiation are all invisible to us.

Furthermore, while we think that we are getting a reasonably detailed image of our surroundings through our eyes, in fact the really good detail is limited to a small area occupying only about two degrees within our visual field. That's about the area of a thumbnail seen at arm's length. Should you doubt this, try looking at one word on this page and then reading other words five lines up or down without moving your eyes. Unless you place the book quite far away, you will find that the detail is too poor even at this slight displacement to read the words.

A likely reason for most of these limitations is that the brain had to evolve with its receiving capacity and processing capacity more or less in step. If the brain received more information than it could use, this would have no immediate survival value and might prove detrimental. Putting it more strongly, many of the severe restrictions on what the human brain receives may well have survival value, because instead of prompting the brain to treat lots of incoming information superficially, they position it to perform incredibly elaborate analyses on a relatively small incoming flow of data.

To get a general sense of how this analysis is done, it is worth briefly reviewing the brain's structure (Figure 5.1). Much of the analysis is done by the cerebral cortex, the intricately folded "gray matter" at the surface of the cerebral hemispheres that consists mostly of nerve cell bodies and receptor branches called "dendrites." However, if you slice below this cortex (as shown in Figure 5.1C), you will find that much of the brain matter underneath is white. This "white matter" consists mainly of the nerve cells' long axons, which serve as transmission cables, and of specialized insulating cells that wrap them in white myelin sheaths.

Beneath this white matter, toward the center and bottom of the cerebrum, is a group of bizarre-looking structures (Figure 5.2). Each of these structures has its own bodies ("nuclei") of gray matter, and some have many such nuclei. These structures have odd double shapes, one half being on each side of the brain, and also have odd names—like hippocampus, amygdala, corpus striatum, hypothalamus, and thalamus. These names used to concern only brain scientists, because we knew little about what the named structures actually did. But we are now becoming aware of what they do, and much of that is important. So, increasingly, these names are working their way into the public eye.

At the center and bottom of the cerebral hemispheres sits the most massive of these cerebral substructures, which is called the thalamus. Under the thalamus, and extending downward to the spinal cord, is a large structure outside the cerebrum called the brain stem, which performs a wide range of basic functions, including control of our conscious states. And finally, tucked behind the brain stem and under the cerebrum is the cerebellum or "little brain," a power-packed computing center whose surface is more deeply furrowed than the cerebrum's and whose form holds as many neurons as all the rest of the brain combined.

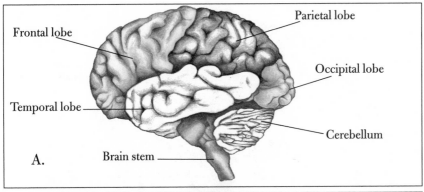

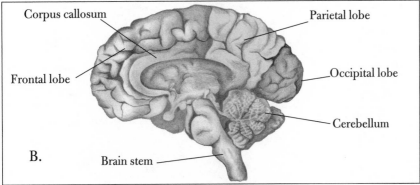

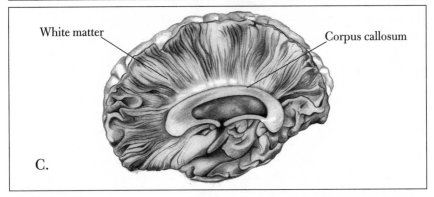

FIGURE 5.1   The human brain. (A) View of the left side of the brain showing the cerebellum, brain stem, and major lobes (occipital, parietal, temporal, and frontal) of the cerebrum. The cerebrum's prefrontal lobes (not labeled in the figure) consist of specific, generally forward parts of the frontal lobes that receive communications from the upper middle (dorsomedial) nucleus of the thalamus. (B) A section through the middle of the brain (slicing through the corpus callosum, cerebellum, and brain stem) that again shows the major lobes of the cerebral cortex, this time on the right side of the brain. (C) The right side of the brain with most of the cerebral cortex cut away to reveal the white matter fibers connecting cortical areas to other parts of the brain.

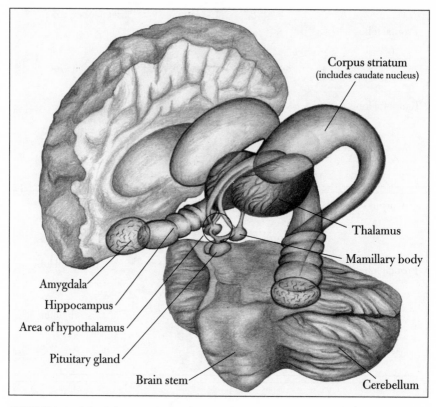

FIGURE 5.2   The thalamus and other deep cerebral structures including the corpus striatum, hippocampus, amygdala, and hypothalamus. For clarity, the structures are drawn in a schematic and simplified form. The hypothalamus (not drawn, but located by the circle in the figure) is a cluster of many small gray matter nuclei. The nature of several features not mentioned elsewhere in the text is as follows: The fornix is a neuronal pathway leading from the two hippocampi to the mammillary bodies and other structures. Each mammillary body (involved with memory, emotion, and motivation) receives information from the hippocampus and projects to the thalamus and other regions. The pituitary gland, which receives instructions from the hypothalamus, regulates the body's other endocrine glands (including the thyroid, gonads, and adrenals) and so plays a critical role in managing vital bodily functions and also in the processes of growth, maturation, and reproduction.

Something quite remarkable about all of these structures—something that has a direct bearing on consciousness—is the fact that they are all highly specialized and subdivided. We can see this in the furrowed gray matter of the cerebral cortex, where specific large regions are mainly devoted to specific tasks like vision, language, touch, movement, thought, and so forth. Beyond that, each of the specialized areas is broken down into an array of small specialty shops—like some vast medieval art studio where thousands of artists collectively produce a mural—some making rough outlines of particular features in pencil, others filling in parts of the background, adding bits of clothing, or painting certain faces. In a similar manner, visual information about an object that you see is relayed to a particular part of the occipital lobe, a part of the cerebral cortex at the back of the brain that deals with vision. From there the information goes to a multitude of specialty shops dealing with things like distance, orientation, shape, color, and motion, and the output from these shops is fed back to the original receiving area as well as forward to other areas, to a point where the object is identified through comparison with existing memory and its relevant features become clear.

Other parts of the cerebral cortex are similarly specialized; but the process does not end there, for in fact the *whole brain* is specialized. The cerebrum's central structures, each finely subdivided in various ways, deal with specific tasks. The hippocampus deals mainly with memory and its consolidation, the amygdala with fear and anxiety, and the hypothalamus with instinctual functions including sleep, hunger, and the sex drive. Likewise, the brain stem has a variety of specific tasks such as communicating or blocking communication with the spinal cord, relaying a torrent of messages from the cerebrum to the cerebellum, regulating the brain in general ways, and governing various brain states including those involved in dreaming, sleeping, and waking. Finally, the cerebellum deals mainly with fine-tuned error correction and coordination, particular parts of its surface (the cerebellar "cortex") being concerned with specific things like coordination of particular kinds of motor acts or coordination of activities relating to "cognition" (perception, memory, and thought).

This remarkable specialization is key to understanding how the brain performs all sorts of mental tasks. For the existence of specific brain areas specializing in things like fear, vision, language, and many other functions means that we can home in on those areas, assess their nature, explore their coordination with other brain regions, and in due course discover how they work. We can also relate malfunctions in different areas to various mental problems; and so, as will be seen in later chapters, this specialization has major implications for understanding and treating mental ills.

Meanwhile, with respect to consciousness, such extensive specialization suggests that there might be a specialized area in the brain that helps to co-ordinate awareness, a "seat of consciousness" if you will. However, this would not be some sort of homunculus or "little man" within the brain looking at what the brain has found. Instead, it would be a data manage-ment center where findings from many brain regions would come together for resolution, where such resolution could cause attention to be focused on certain matters or redirected, and where messages would be issued to other brain regions as a result—messages engaging them to perform various kinds of further analysis or action.

In the early days of brain science, many experts tended to shrug their shoulders and say that consciousness is "distributed" about the brain, that when you get enough neurons together it just "happens." But there is a problem here, because the brain is so compartmentalized, with different brain areas doing different things. And your brain is not like a computer, where impulses zip around at the speed of light—186,000 miles (seven-and-a-half times around the world) in a second. Instead, brain messages plod along relatively slowly, typically at somewhere between walking speed (as slow as 3 miles per hour) in unsheathed dendrites and jet takeoff speed (about 200 miles per hour) in well-sheathed axons.

For this reason, the brain tries to avoid processing one thing at a time like a computer. Instead, it is what we call a "parallel processor," handling thou-sands or millions of impulses at the same time. This means that if one brain area wants to send a complex message to another far-flung area, it needs to send a lot of information at one go. If the receiving area is not nearby, this requires a substantial neuron cable. And this suggests that if many far-flung areas had to talk back and forth in a marginally coordinated fashion to bring things together in consciousness, then your brain would consist of cables and little else.

But if the brain has a consciousness coordination center, where is it? The most obvious place to look is in the cerebral cortex, where so many things (touch, sight, language, thought, and motor acts) that relate to human consciousness get processed. But we can pretty much rule the cerebral cortex out. To begin with, it is the most accessible brain region, and we know too much about it. We know what its various specialized regions do in some detail, and none of them seems to handle something as complex as consciousness coordination. Also, since consciousness involves many brain functions and nerve impulses travel slowly, any coordinating center should presumably be situated toward the middle of the brain for easy access, rather than being where most of the cerebral cortex is—on the brain's outer fringes. In addition, as already noted, infants appear to be conscious, even though large parts of their cerebral cortex are immature; and a rich record of all sorts of brain injuries fails to indicate that damage to any particular part of the cerebral cortex destroys consciousness.

A more productive approach might be to list the features that a consciousness coordination center should have, and then see if any particular brain structure matches up. Besides having a central location (ironically, near Descartes' now-ignored pineal body), such a center should be fairly large (to permit receipt, coordination, and rerouting of many complex inputs). It should have a varied composition (to deal with the many varied types of information being coordinated). And it should receive massive inputs from nearly all parts of the brain that process sensory information or engage in higher thought. And it should be intimately involved with attention, since a major reason for coordinating consciousness is to direct attention. Beyond that, examination of the medical record should indicate that damage to this structure is fatal or else that it harms or destroys consciousness. And finally, the structure's inputs, outputs, or activities should offer some explanation as to why all the things we experience in consciousness seem so well-coordinated with one another.

This matter of coordination is key, because things do get together in consciousness. In fact, they get together superbly. The conscious mind sees a world in which touch, vision, hearing, language, motion, memory, higher thought, attention, speech, and emotion are all included. What's more, the coordination appears seamless. Vision seems smoothly coordinated with hearing and motion, thought with memory and attention. Diverse movements seem smoothly coordinated with each other as well as with sensory inputs, attention, and higher thought. In fact, virtually everything seems minutely coordinated with everything else—a big hurdle for theorists trying to get to the bottom of how consciousness actually works.

A few years back, a flurry of scientific and popular articles highlighted a possible role of the cerebellum in all this. These articles, from various sources, indicated that the cerebellum was not limited to coordinating motor acts, as had once been thought, but that it provided fine-tuned coordination for all sorts of things like memory, vision, spatial orientation, touch, language processing, planning, foresight, judgment, attention, motivation, emotion, and integration of higher-order behavior. To this end, it was said to receive and process a torrent of diverse information coming in from specialized parts of the cerebral cortex and other brain areas.

None of this gives any reason to think that a "seat of consciousness" resides within the cerebellum, which is far from the brain's center, lacks a varied structure, and can be damaged or even absent (in people lacking it from birth) without greatly affecting consciousness. But since consciousness is so well coordinated and the cerebellum appears to be the brain's master coordinator for multitudes of intricate brain activities involved with consciousness, there seems reason to ask where the cerebellum sends its output.

It turns out that most of the cerebellum's output goes back to the cerebral cortex, completing the information loop; but before it gets there it stops off and passes through a synapse in the thalamus. That's very interesting—indeed, it is highly suggestive. We know that the thalamus acts as a relay station, receiving all the incoming information on everything you see, hear, taste, and touch, and sending it out to other parts of the brain. But the thalamus is more than just a relay station—because it receives vast amounts of information back from virtually all parts of the brain; and the massive conduits carrying this processed information back to the thalamus are at least as large as the impressive conduits the thalamus uses to receive and send out sensory information.

The thalamus also has many of the other features we would expect to find in a consciousness coordination center—being large, centrally located, and endowed with a richly varied structure consisting of numerous gray matter nuclei with a wide range of different shapes and sizes.

Not only is the thalamus connected intimately to the cerebral cortex, other cerebral structures, and the cerebellum, but it also serves as an upward extension of the brain stem. It thus appears to be at the center of our neural system with all the structural and functional attributes needed to support and dynamically regulate consciousness.

Furthermore, a variety of experiments and medical case histories have deeply implicated the thalamus in consciousness. To begin with, a good number of coma cases, including the famous case of Karen Ann Quinlan, have convincingly demonstrated that damage to the thalamus can leave someone alive but permanently deprived of consciousness. And while damage to the brain stem can produce similar effects, the brain stem is known to govern the thalamus, and so certain kinds of brain stem damage could also be affecting the thalamus.

On another tack, researchers have looked inside the thalamus in order to explore visual attention. We know that visual areas in the occipital lobe send processed visual information to a substructure in the thalamus containing "retinotopic" maps that in essence recreate the eyes' visual field. This same substructure also has connections to nonvisual parts of the cerebral cortex. When people engaged in a visual task are exposed to distracting influences, they show increased brain activity in this substructure—which is what we would expect if they were increasing their attention to better focus on the task. Also, monkey experiments have shown that chemically inhibiting a small part of this substructure makes it harder for the monkey to shift visual attention, while reducing such inhibition has the opposite effect.

Thus we appear to have a model that does not locate the emergence, coordination, and application of consciousness but that walks us through it. That is, we are not aware of the raw information sent to the brain by the optic nerve. What we become aware of is this visual information after it has been processed by the visual cortex. At least in part, this appears to be the information sent from the visual cortex to the retinotopic maps in the thalamus. And the research cited above shows that activities in this part of the thalamus relate intimately to the job of directing visual attention—work usually deemed to require awareness and to be a conscious task.

All this makes it clear that the thalamus is deeply involved in human consciousness. Not surprisingly, we are neither the first nor the foremost to air this view. The first appears to have been Wilder Penfield, author of *Mystery of the Mind* (Princeton University Press, 1975); another is Joseph Bogen, the neurosurgeon who collaborated with Roger Sperry in his famous "split-brain" experiments; and the foremost is Francis Crick, codiscoverer of DNA, cognitive scientist, and author of *The Astonishing Hypothesis: The Scientific Search for the Soul* (Simon and Schuster, 1994).

Crick's avowedly speculative view is that the visual cortex probably generates the visual awareness that we consider part of consciousness, and that the thalamus uses this to direct attention—and so visual consciousness occurs in the visual cortex. It should be noted, however, that he is observing things from a visual perspective, having concerned himself primarily with visual processing, and this may have influenced his view.

We are coming in from a slightly different direction. We have been drawn to the striking coordination that characterizes consciousness. As noted earlier, the brain loves to specialize, and here is a golden opportunity to do so. For it seems efficient to bring the results of all sorts of mental processing together in one structure, coordinate them there, and broadcast this coordinated information so that the brain's attention can be properly directed and a medley of brain structures can be engaged in activities ranging from football to philosophy. Indeed, this seems far more efficient than trying to coordinate things between various "conscious" parts of the brain without having a coordination center. And while none of the things we have learned about the thalamus, its visual centers, and its relation to the cerebellum tells us what is really happening, they lend credence to the idea that the thalamus is coordinating elements of consciousness received from different places in the brain.

Would that make the thalamus a kind of homunculus or "little man"? It would not, for the thalamus, like consciousness, is very limited. On its own, as we see it, the thalamus is just a coordinator. Like an orchestra conductor, it has no instrument of its own to play. It cannot see, hear, have feelings, think, or even exercise free will. But it receives vast ongoing cascades of nerve impulses from brain areas that do all that and more. Its connections suggest that it can coordinate these cascades, orchestrate attention, summon memories, direct thought, order up all sorts of analysis, receive fresh cascades of nerve impulses reporting the results of such analysis, and tell the brain's premotor and motor areas to have the body perform all sorts of "conscious" acts. If it actually does all this, then the thalamus, despite its limitations, would appear to have superb across-the-board access to present and potential outputs from virtually all parts of the brain, and would appear to be doing much of what we would expect a coordinator of consciousness to do.

That's exciting. But if we are ever going to unravel the master magician's sleight-of-hand, we need to take a more dynamic view. For, as William James, the father of American psychology, pointed out over a century ago, consciousness is not a structure. Consciousness is a process. It is a process like your arm's act of throwing a ball or the descent of water over a falls. It makes no sense to look for a "center of throwing" in your arm, because nearly all parts of the arm, as well as many other parts of the body, are involved; nor can the essence of a waterfall be found in the responsible force of gravity, or in the water, or in the rocks the water strikes. So even as we come to see that the thalamus almost certainly plays a key role in consciousness, we need to take our explorations beyond structure and examine consciousness as a process.

How can we do that? A good way to start is to look at the brain's love of what some authors call "loops." These aren't really loops; what they are is complete neural circuits. But whatever one calls them, they seem to be everywhere within the brain. For instance, as we have seen, lots of areas in the cerebral cortex send information to the cerebellum, which processes it and sends it to the thalamus, which sends it back to the originating areas in the cerebral cortex, thereby completing a multitude of circuits. Or centers in the cerebral cortex send information out to nearby cortical areas for processing and have the results fed back to them to complete innumerable loops. Or various brain areas, including vast arrays of them in the cerebral cortex, receive information from the thalamus, process it, send it back to the thalamus, and get it back in turn. The possibilities seem endless, and the realities are practically endless; for the brain indeed loves loops and fills itself with them.

These loops have a strong potential for repeating, maintaining, coordinating, amending, and refining information associated with all sorts of brain processes, foremost among these being consciousness. What's more, the timing looks right for them to make an essential contribution to consciousness. As we have seen, brain impulses travel slowly. But the circuits we are talking about are fairly short. For example, consider an imaginary loop running from the thalamus to the cerebral cortex and back, a loop in what is known as the "thalamocortical system" that takes in both the thalamus and the cerebral cortex. If the loop's complete circuit is something like 9 inches long, which seems reasonable, then a series of nerve impulses traveling around this loop at 60 miles an hour (88 feet per second) could make 12 complete circuits in a tenth of a second. Since we cannot make a single conscious decision in less than a few tenths of a second, an active circuit of this kind would seem able to serve consciousness. And so far as we can tell, this is precisely what vast numbers of circuits within the thalamocortical system actually do.

Some such vehicle is clearly needed, because experimental studies have shown that single conscious decisions can call forth essentially simultaneous activity in many parts of the brain. So even if the thalamus acts as a central coordinator and gating system, the brain needs ways of circulating masses of information between the thalamus and various processing areas in a speedy and recurrent fashion, so that all of the areas involved can work together.

This business of working together is important, not just for timing but for other reasons. As pointed out in a powerful theory of consciousness advanced by noted cognitive scientists Giulio Tononi and Gerald Edelman, it looks like *all* the circuits active in consciousness at any time are highly integrated with each other, so they all work together as a unit.* Common sense

---

*As will be seen in the next chapter, the general character of this integration is influenced by brain stem mechanisms that control the activation, input–output gating, and chemical modulation of our thalamocortical networks.

FIGURE 5.3   A goblet or two faces, depending on one's momentary point of view.

supports this view. For anyone can see that we tend to do conscious things one at a time and have trouble doing two or more together—unless one of them is an automatic act (like walking or chewing gum) that requires little or no conscious attention. As an example, when you look at the drawings in Figure 5.3, you can see them as two faces (black) or a goblet (white), and your chosen view can change from one moment to the next. What you cannot do easily is see either drawing as both faces and a goblet simultaneously. This suggests that the billions of neurons creating your sense of consciousness are not acting independently but instead are all working together as a unit.

Further supporting this idea, computer simulations have shown such large-scale integration to be feasible. That is, if simulated neuron circuits fire in a continuing, recurrent, and highly parallel fashion, they can "bind" the activities of different specialized neuron groups, making them operate together as a unit. Of course, this dynamic integration depends on continuing stimulation. So neuron groups that don't get these ongoing, recurrent, parallel signals will stop contributing to consciousness; but the signaling neurons can also reach out to other groups not currently contributing to consciousness and bring them in. For example, if you were looking at, say, a yellow bird, numerous visual processing circuits would be among those bound together; but if the bird flew off and you then tried to recite a poem about a yellow bird from memory, visual involvement would diminish, and the predominant bound circuits would be ones dealing with memory of language and with speech. This suggests that while our integrated consciousness system is probably "anchored" in the thalamus (the prime coordinator and gating mechanism), the system can probably rove about among specialized neuron groups in the cerebral cortex and elsewhere, and so its location in the brain can shift with the task at hand.

So what happens if we prevent half the brain's neurons in the cerebral cortex from signaling to the other half in a quick, recurrent, parallel fashion? We can do this by surgically severing the corpus callosum, a thick bridge of neural connections that joins the two cerebral hemispheres. In fact this has been done to help patients with severe epilepsy, and study of such "split-brain" subjects points to an astounding answer: The patient's single integrated consciousness splits into two distinct systems, each one operating as an integrated unit.

This conclusion first emerged from animal research that Roger Sperry and colleagues did in the 1950s, for which Sperry later received the Nobel Prize. Through carefully designed experiments, these researchers found that a cat or monkey whose corpus callosum had been severed could be trained in such a way that the right hemisphere would receive different training information than the left and would also respond differently. As Sperry explained, "It is as if the animal had two separate brains."[1]

Following up on this, a team led by Sperry and Michael Gazzaniga began working with a charming, take-charge World War II veteran now known in the literature by his initials, W.J., whose corpus callosum had been severed in an attempt to quell frequent epileptic fits. The scientists knew that the left cerebral hemisphere generally controls speech and right-side functions (including the right hand and information received through the eyes' right visual field), while the right hemisphere controls left-side functions. Therefore, they flashed images (bursts of light, boldface letters, and so on) upon the left or right side of a screen for enough time to register in W.J.'s left or right visual field, but not enough time for him to move his head and register the image in both visual fields.

The results were striking. When the images were flashed on the right side of the screen (reaching W.J.'s left hemisphere with its control over speech), he was able to identify the images with no trouble and tell the investigators what he saw. But when they switched the images to the left side of the screen, it was as though he had gone blind, because he claimed insistently that he could not see anything flashed on the screen.

But of course this was his left hemisphere speaking. The researchers had not expected that his left hemisphere would see anything (not if the left-screen flashes registered only with the right hemisphere), so they had given the right hemisphere a way to respond. That is, they had put his left hand (controlled by the right hemisphere) on a telegraph key and had instructed W.J. to press the key whenever he saw an image. And they found that the telegraph key was pressed repeatedly in response to each image flashed on the screen, even as W.J.'s voice (controlled by the left hemisphere) was denying that he saw anything at all.

Despite these bizarre results, W.J. and most other split-brain subjects appear outwardly normal, because their two hemispheres are receiving similar sensory inputs; the two hemispheres are also good at cuing one another with hand motions and like devices to create an impression of unity; and the brain is in fact connected elsewhere—enough to permit coordination of some functions but not integration of the myriad thalamocortical circuits.

Of course, these initial experiments created quite a stir; and since then Gazzaniga and others have developed elegant tools that allow them to send more elaborate messages to the right or left hemisphere of split-brain subjects. One sort of experiment conducted with such tools has a direct bearing on the concept of integrated consciousness. That is, when a split-brain subject's two hemispheres are given two separate problems that could be seen as part of a more complex larger problem, each hemisphere solves its own small problem independently and the two sides give two independent answers. However, a person whose brain has not been split responds differently. Upon receiving the same two problems through separate instructions to the right and left hemispheres, a normal person proceeds to lump the problems together, solve the much more difficult larger problem, and provide a single answer. This is strong evidence that consciousness does in fact operate as a single integrated unit unless the conditions needed for integration disappear.

But in Tononi and Edelman's view, which we tend to support, integration is not enough. For consciousness is indeed like a waterfall. The essence of any waterfall arises from the diversity imposed upon the falling water by the rocks. Theoretically, all the water could be sucked into a pipe draining the stream above the falls and pumped over the falls in a mighty jet that doesn't touch the rocks. In that case the water would still be "integrated," but the diversity imposed by the rocks—and hence all of the falls' natural beauty—would be gone.

In a similar fashion, consciousness appears to depend upon diversity—upon having the integrated consciousness system engage innumerable bits of incoming information, bits that are digested and used to generate a quick progression of new conscious scenes (like the frames in a movie film) whose content is limited only by imagination, experience, and the stream of incoming signals. Thus, the key thing is to introduce a veritable Niagara of diversity within a system that remains synchronized, cohesive, and operating as an integrated whole.

We can clearly see diversity's role by looking at what loss of diversity does to consciousness. Clearly, diversity is lost during an epileptic seizure, when vast populations of neurons throughout the brain are all firing together in regular synchronous bursts that are like the water-jet—that is, highly integrated but without the diversity needed to produce a waterfall. It thus comes as no surprise to find that during such a seizure the person having it is unconscious. Or take a more routine example that happens to us all every night. If the thalamocortical system gets little news from the outside and is left pretty much to itself, as it is when we start to sleep, its cells begin firing in regular synchronous bursts. As a result, we fall into a deep slumber, and what we perceive as consciousness is lost.

In the latter case, consciousness is restored by the hypothalamus, the brain stem, and a strange layer of gray matter called the reticular nucleus of the thalamus that wraps around the rest of the thalamus like the skin of a grapefruit. Nearly all the neurons that pass between the rest of the thalamus and the cerebral cortex go through this reticular nucleus. And as the brain gets ready to awaken or to enter the phase of dreaming slumber called REM (rapid eye movement) sleep, the hypothalamus sends instructions to the brain stem; the brain stem activates the reticular nucleus; and the activated reticular nucleus prompts many of the neurons passing through it to inhibit their neighbors. This disrupts the synchronized burst firing pattern, enabling new data to enter the system and creating enough diversity in the integrated system to restore consciousness—either the strange consciousness of the dream state or the more "normal" consciousness that we enjoy when we're awake.

This apparent raising of consciousness by integrated diversity could explain why symphonies, sunsets, and waterfalls all have such appeal. For these things are both highly integrated and highly diverse. So they could be doing exactly what we feel them to be doing—stimulating, enriching, and enlarging our sense of consciousness by providing a large flood of information that, like consciousness itself, is both integrated and diverse.

Be this as it may, it seems clear that our new brain science insights are shedding significant light on human consciousness and the nature of mental ills. To begin with, they are demonstrating that the thalamus is deeply involved with human consciousness; and since all mental ills involve the conscious mind, we should not be surprised to see the thalamus playing a role in various mental ills, or to find that schizophrenia research (to be discussed in Chapter 10) has implicated the thalamus as one of the key players involved in that disorder.

We can also see that a properly working sense of consciousness requires precise coordination by a lot of different brain centers. That makes coordination between brain centers a major issue and highlights its role in mental ills. Of course, trouble in any of a multitude of brain centers can cause illness. But most illnesses that can be traced to one brain area (illnesses caused by things such as strokes, tumors, or physical damage) are considered "organic" and are assigned to neurology. Therefore, we suspect that many of the mental ills remaining within psychiatry's domain seem mysterious because they arise from troubled coordination between multiple brain centers.

Beyond that, our original idea that consciousness is limited seems strongly supported by what we know of the brain's organization, specialized nature, and style of operation. This implies that there is lots of unconscious activity within the brain, but also that most of this unconscious activity is highly specialized. These points, clearly relevant to both psychology and pharmacology, will be explored more fully later.

More broadly, this brief review of consciousness and basic facts about the brain shows why we believe brain science is well-positioned to bring psychology, therapy, and biomedicine together. For even in this general field of consciousness, one traditionally reserved for philosophers and other humanists, it seems clear that brain science is making progress—that it has lifted us out of our audience seats and taken us backstage to a place where we can see how the master magician does his work. No more must we treat the mind as an abstraction or the brain as a black box. For in fact, animal experiments, imaging techniques, and other methods have allowed us to see inside the brain and examine how it works.

Of course, with its 100 billion neurons the brain is a complex place, at least as complex for the average scientist as New York City is for the average tourist. So we have no reason to expect full revelation. But we have already seen enough to get an idea of what is happening, to dispel the idea that the mind is somehow beyond the reach of brain science, to demonstrate that the brain and mind are different sides of the same thing, and to conclude that much as a computer's innards produce the output seen upon its screen, so the brain's far vaster operations produce what we experience as activities of the conscious mind.

6

Sleep and the Dance of Dreams

Let the dreamer awake and you will see psychosis.

—Carl Jung

Moving away from daytime consciousness and one step closer to mental ills, we come to the land of sleep and dreams. Here the relevance to psychiatry is clearer. As in Alice Morrisey's case, many mental patients have disturbed sleep patterns, and dreams provide a strange view of the mind's world—one so dramatic and bizarre that not only psychoanalysts but untold generations of prophets and soothsayers have dedicated themselves to divining what dreams mean.

But neither sleep nor dreams have yielded up their secrets easily. In contrast to someone who is awake, of course, a sleeping person can't be questioned. So in the past philosophers, psychologists, and other students of the mind tended to regard sleep as a closed book. The only pages open to them were those that dealt with dreams. And while dreams could be fascinating, they seemed so strange that dream interpretation became an iffy business reserved to seasoned professionals of one sort or another. So until fairly recently, most people regarded sleep as a period of brain rest and dreams as a baffling mystery.

But as twentieth-century brain science emerged, it found that the sleep book could be opened. Brain scientists don't require conversations with their subjects to gather data. What's more, the brains of experimental animals are sometimes easier to study when the animals are asleep, and the same is true of the brains of human volunteers in sleep laboratories. So scientists did probe sleep. They explored the brain anatomy and brain chemistry of sleep. They recorded sleepers' brain waves and awoke them at particular sleep stages to see what they were dreaming. They learned volumes about both sleep and dreams, and they developed powerful and compelling theories about what happens during sleep, how sleep relates to dreams, how brain chemistry generates the dream state, and why dreams are so peculiar.

This flood of scientific information is useful to psychiatry. It shows how dreams should be interpreted (not the way the Freudians do it) and how they can improve our understanding of problems troubling our patients. It lays bare simple chemical mechanisms responsible for the sleep disorders that assail multitudes of people, including millions with depression and other mental ills. It reveals that brain chemistry alterations associated with sleep and dreams resemble alterations that produce the active symptoms of many mental ills. And most startling of all, it shows how nearly everybody experiences brain chemistry changes in their sleep that make them both conscious and psychotic every night—thereby providing a model that is useful for understanding psychotic states, counseling mental patients, reducing the stigma of mental ills, and coming to grips with the true powers and pitfalls of psychiatric drugs.

Let's begin with the fact that sleep is vital. This idea seems reasonable on evolutionary grounds alone, because a sleeping creature runs a relatively high risk of being attacked and eaten. That cuts its chances for survival. And while some reptiles, birds, and mammals get less sleep than others, we know of none that foregoes sleep altogether. So there must be good evolutionary reasons for paying the price of sleep.

This theory is strongly supported by experimental evidence. People deprived of sleep perceive that they are becoming less effective, and eventually they begin having hallucinations and other mental problems. Beyond that, a rat deprived utterly of sleep will die in about a month of various brain-related malfunctions (defective regulation of body temperature, dietary energy flow, and the immune system), whereas a companion rat that is not so deprived will remain alive and well.

Surprisingly, however, we have not yet penetrated to the heart of sleep's role in these protective bodily functions; nor have we fully solved the mystery surrounding other sleep-related benefits. But we have plenty of compelling theories, and in fact we are beginning to understand sleep's contribution to important mental functions. In particular, work by many researchers has shown sleep to be deeply involved with learning and memory. So it seems likely that the sleeping brain is busily sorting out the day's information, associating it with other related information in its vast stores, and blending these bodies of data in order to be coherently prepared for the future. Also, it seems clear that sleep must have a restorative function, and that processes which happen during sleep are needed to keep the brain attuned and working well.

What we have learned about sleep, and it is considerable, we owe largely to sleep research. The beginnings of such research hark back to 1928, when Hans Berger, a German psychiatrist, placed electrodes on the scalp of a human subject and recorded continuous electrical activity. What convinced Berger that his first "electroencephalograms" (EEGs) truly reflected brain activity rather than body movement or something else was the fact that the patterns of electrical activity changed when his subjects slept.

We now know what caused this electrical activity and its changing patterns. An active neuron can fire up to 100 times a second. This firing involves a small amount of electrical activity, and the firing of billions of neurons in one part of the brain places a small but detectable electric charge on overlying portions of the scalp. The charge varies from place to place on the scalp and also varies in each small fraction of a second according to local neurons' firing patterns. So two electrodes placed on different parts of the scalp will register two sets of fluctuating charges. If the two electrodes are hooked together and attached to a recorder-amplifier, the device will show a small but rapidly changing voltage that can be charted as "brain waves" on an oscilloscope or moving roll of paper.

When the subject is awake, the EEG typically records a high frequency (rapidly fluctuating current) and low voltage (small peaks and troughs). The high frequency probably reflects diverse activity by many neurons, while the low voltage suggests that the number of neurons firing together in a highly rhythmical or fully synchronized manner is fairly small. However, as the subject goes to sleep and descends into progressively deeper sleep stages (from what is known as Stage I to Stage IV sleep), this pattern changes. The frequency falls, suggesting less diverse neuron activity; but the voltage rises, creating higher peaks and deeper troughs. This indicates that more of the neurons are firing together in a synchronized manner and that consciousness has been seriously compromised or lost.

After a gradual and steady descent from waking to Stage IV, the sleep cycle changes direction and the sleeper returns quickly to Stage I. For any normal sleeper, this whole cycle repeats itself about every hour and a half (90 to 100 minutes) through the night, until the sleeper is awakened by a laboratory technician—or by the familiar panoply of dawn, alarm clocks, noise, the internal urgings of daily (circadian) rhythms, and the complex restorative effects of sleep.

None of these findings from the early EEG days dispelled the idea that the sleeping brain was resting. What shattered that idea, over twenty years after the EEG's emergence, was revelations about Stage I. In 1953 Eugene Aserinsky, a researcher investigating eye movements, found that spates of rapid eye movement (now called REM) occurred in Stage I, accompanied by increased respiration and a faster heartbeat. A closer look at this REM-associated Stage I sleep pointed to high levels of brain activity. Indeed, in many ways the brain seemed as active in REM sleep as when it was awake.

Also, REM sleep seemed to host most dreams, because sleepers awakened from REM sleep were many times more likely to report a dream than those awakened from non-REM sleep. Furthermore, REM sleep accounted for nearly all the colorful and dramatic dreams, with the dreams reported after non-REM sleep tending to be highly repetitive and dull.

As often happens, these discoveries raised more questions than they answered. For instance, What caused the REM state? If the brain was so active in REM sleep, why did it just stimulate the heart, lungs, and eyes? Why didn't it produce bodily movement? Why wasn't the sleeping brain awake? And how was the REM state tied to the ancient mystery of dreams?

Of all these questions, the last was the most compelling. Throughout history a vast assortment of mystics, priests, philosophers, and seers have tried to explain our surreal and dramatic dreams with scant success. Indeed, for most of the past century the best we could do was agree or disagree with the old (circa 1900) dream theory of Sigmund Freud, who claimed that dreams arose from a troubled subconscious, contained forbidden thoughts we couldn't bear to acknowledge, and seemed bizarre or choppy because a mental censor disguised or removed the forbidden thoughts.

Like Freud's related beliefs about repression, childhood sexuality, and ego psychology, his dream theory was hard to prove. So, like most other early tenets of psychoanalysis, it attracted less and less attention as the Freudian zenith passed, and despite periodic attempts at revival, it now seems bound for the land of dreams itself. But Freud was an astute observer, one who had lots of experience with dreams. He could see that dreams were bizarre and choppy, sometimes involved old memories, tended to be emotion-laden, and often dealt with troubles that concerned his patients. Because he had no other choice, his theory was based on guesswork; but even so, he saw a clear need to have his theory explain these prominent dream features, and so it did.

We can see some of the dream features Freud observed by examining the reports of volunteers asked to describe their dreams upon awaking. For instance, consider the following account:

> I am in the mansion of a very wealthy woman who is my mentor or benefactor. I am to work for her as a nanny for her two children. She shows me around the house as she describes the children's routine. First, they go to the pool for a swim. Then they take a bath, one at a time. Then I help them dress and feed them, etc. The house was fantastic, especially the bathtub. It is up in a loft with a low ceiling. The tub was light blue and kidney shaped. The room had a soft blue glow that was very relaxing. I got in the tub as she was talking, hoping that she

wouldn't notice. As I looked at the water, I noticed that it was dirty. I got out and we continued our tour. (I was bone dry when I got out.) I worried about how I would entertain the children (ages 7 and 8), and knew it would be a hard job.

It is the next day and I am in charge. The woman is out of town. We go the pool, which is now a beach. I wash the sand off the kids' feet. I look up and notice how attractive the boy is. He looks at me provocatively. At first I am bewildered. Suddenly, I am filled with lust for the boy. I turn to the girl and wash the sand off her feet. We return to the house. My mother is there. I must have invited her there to show her my good fortune at living in such grand style. I point out all the nice antiques, and she is impressed and happy for me. I go back to the house but don't know where I had been. I am very anxious to see the boy again. I am getting him ready for his bath. In the background I hear some women whispering about how I was acting improperly with the boy.

This particular dream happens to contain certain elements that Freud singled out for attention—most notably sexual feelings and parental relationships. But these dream elements are not disguised as Freud would assume. And not all dreams are like this; in fact, it is very common to find dreams that involve strong feelings but that lack any obvious parental or erotic content. Here is another dream reported by the same woman:

I am working in a restaurant. It is the end of the night and I can't wait until we close. I am trying to clean up and need to empty the trash. There is a double bag with hot oil in it, and I only pick up one because I am so tired. Immediately, oil starts leaking all over the floor. I am frantically looking for somewhere to put it. It becomes a burlap sack and is leaking on my legs, but the hot oil doesn't burn me. Finally, I get rid of it and no one notices. I am desperately hoping that no one slips on the oil, but I can't seem to clean it up.

Two customers come in. A manager tells them that we are closed. The man begs for a bowl of fish chowder. I take the man's order for chowder, wine, and orange juice reluctantly. After much deliberation I bring the drinks. Now he tells me he wants the chowder to go. I try to reach a container over the take-out counter, then realize that I have to go around. It is taking forever. As I am putting the lid on the container, I see that it is only half full. I reach for a large white vase to fill it with, but it is vinegar that pours out. I am filled with dread that I have to start over.

Obviously, anyone eager to find erotic or parental influences in this second dream could do so by assigning symbolic meanings to items in the dream. But that assumes that symbolism plays a big role in dreams, and we really have no evidence of that. Nor, as already noted, do we have any convincing evidence to support Freud's view that dreams are edited by a hidden censor.

Of course, if we abandon both of these devices, that leaves a lot to explain. For instance, if our vivid dream illusions don't arise whole in the subconscious, how do they arise? Why do they register in the dreamer's conscious mind? Why are they so bizarre and disjointed? Why are they commonly charged with emotion? Why does the dreamer accept the dream and all its impossible situations as being valid? And why, after entering another sleep stage or awaking, does the dreamer typically forget the whole thing?

In 1968, when I (Allan Hobson) set up a small neurophysiology laboratory at the Massachusetts Mental Health Center with the help of Robert McCarley, the answers to these questions were unclear. That didn't bother us, because we didn't set out to study dreams. We set out to map the cat brain stem, probing brain stems with microelectrodes small and sensitive enough to record the firing of individual neurons. That satisfied the conditions of our grant and paid the bills. It also allowed us to explore the interesting question of how the brain stem related to REM sleep.

It turns out that cats are well-suited to such research, because their sleep patterns resemble those of humans. In fact the brain waves, rapid eye movements, and muscle twitches exhibited by cats clearly distinguish periods of REM sleep from non-REM sleep very much the way they do in humans.

We knew that the REM process doesn't start with the brain stem. Instead it starts with the hypothalamus, a key brain structure near the base of the cerebrum containing many small gray matter "nuclei" that deal with a wide range of body functions. One of these sits astride the optic "chiasm" (the point where the optic nerves from the two eyes cross). This small dab of tissue contains the brain's mighty circadian (24-hour) clock. It can reset the clock to fit light-related sensory data coming in from the optic nerves, thereby adjusting for jet travel and other factors; and it also helps govern all sorts of brain and body activities that follow the clock's daily cycle, among other things by sending periodic signals to the brain stem directing that REM sleep be turned on.

Unfortunately, it's hard to explore the hypothalamus with microelectrodes, because it contains many small distinct nuclei doing different things, and because many of the neurons in these nuclei are very small. So we and other REM sleep researchers turned to the brain stem, the next structure in the chain of command, which we knew to play a major role in REM and which was easier to probe.

By the early 1970s it had become clear that the brain stem was a major source of serotonin and norepinephrine, two of the brain's key neuromodulators. These two chemicals govern the activities of hordes of neurons, play crucial roles in a number of brain functions (including consciousness, remembering, thought, and judgment), and help define the entire brain's mental state. But norepinephrine and serotonin did not appear to be major players in REM sleep. Instead another modulator, acetylcholine, was known to circulate actively through the brain in REM sleep. Therefore, we wanted to learn more about how the brain stem, REM sleep, and these three neuromodulators interacted.

One thing we already knew (something basic to our story) was the critically important difference between the brain's neuromodulators and its more basic neurotransmitters. Of the two, the neurotransmitters are easier to understand. Most discharging neurons release one of two such transmitters into the synapses at the end of their axons—these being either glutamate, which encourages a receptor neuron to fire faster, or GABA (short for gamma-aminobutyric acid), which prompts it to fire more slowly. The main point here is that these transmitters tend to act mainly as simple messengers—also known as "first" messengers. And while these first messengers, like the god Mercury, can prove very influential, their effects are generally brief.

Not so the three neuromodulators we were watching. The brain has lots of different modulators; but these three and a fourth one, dopamine, are clearly major players. Like the simple transmitters, these modulators get released by neurons, enter synapses, and find docking sites on receiving neurons. But they don't urge the receiving cells to fire directly. Instead, as noted earlier, they cause chemical changes inside the receiving neurons that increase or reduce their sensitivity to the basic messengers—much as the words "READY, SET . . ." prepare a sprinter for the starter's gun, or the sight of a blinking light at an approaching intersection prepares a driver to slow down.

These changes inside the cell can be long-lasting. Indeed, under some circumstances they can be permanent. So while the first messengers typically do their reversible work in a brief fraction of a second, the slower-acting neuromodulators produce changes that can last for minutes, hours, months, years, or a lifetime.

Brain scientists sorting all this out have become fond of referring to "second messenger" and "third messenger" effects. "Second messenger" effects alter a receiving neuron's internal chemistry and surface membrane without targeting its nucleus directly. In contrast, "third messenger" effects enlist the power of the nucleus and its genetic material to make changes.

Whatever the precise methods used, these modulators have vast power. For they don't just tweak a few neurons. They influence *billions* of neurons. And by making the affected neurons more or less ready to fire and interact with others, they can activate or deactivate whole sections of the brain, turn consciousness on or off, and determine how all sorts of other brain functions are performed. So they are not really like isolated footrace starters or blinking lights. They are more akin to awesome and elemental forces of Nature—rather like the trade winds that bring gentle rain for crops when they're blowing right, but that also have the power to create hurricanes, droughts, and floods when things go wrong.

One day in 1973 a researcher named Peter Wyzinski was exploring a brain stem in our neurophysiology lab when he found something very interesting. What he found was a neuron that failed to become more active as REM sleep started. Instead it became less active. In fact it reduced its firing rate to just about zero and remained essentially silent for much of the REM period, before picking up its firing rate as REM sleep ended. Realizing this could be important, we kept monitoring the cell for several REM cycles to confirm we had correctly observed what it was doing.

We later found that Wyzinski's microelectrode had been a hair off the mark. Instead of probing its intended target—a part of the brain stem that turns on in REM sleep—it had been in a place called the locus coeruleus, a key center for the distribution of norepinephrine, which is taken by neurons in this area and sent through their long axons to other parts of the brain. Further research showed that brain stem neurons responsible for providing the brain with most of its norepinephrine, and also with serotonin, turned off in REM sleep.

Working with this discovery, Robert McCarley and I became convinced that a simple mechanism in the brain stem could account for REM sleep. What seemed to be happening was this: When a person who is awake goes to sleep, external signals to the brain diminish (lights are switched off, the person's eyes close, movement is reduced), circadian signals coming from the hypothalamus change, and the brain's general level of activity falls. As this happens, various groups of brain stem neurons—including those that supply the brain with norepinephrine and serotonin—reduce their own levels of activity because they are inhibited. So the circulating levels of norepinephrine and serotonin in the brain stem and elsewhere fall to an intermediate point as sleep proceeds through its four stages.

But suddenly, around the end of Stage IV sleep, these levels fall precipitously. In essence, the brain stem cells supplying these two modulators turn off altogether. (We now know that the hypothalamus issues a "stop work" order to brain stem neurons in a center called the raphe nucleus, which supplies the brain with most of its serotonin.) This shutdown causes a drastic shortage of norepinephrine and serotonin, and the serotonin shortfall unleashes previously repressed neurons responsible for supplying the brain with acetylcholine. These latter release acetylcholine to the brain like a genie released from its lamp; and the resulting high acetylcholine levels excite visual, motor, emotional, and certain other centers; raise the sleeping brain's general level of arousal; and raise or restore consciousness.

Like our sleeper's brain, the brain of someone who is awake also has significant levels of circulating acetylcholine; but serotonin and norepinephrine are present then, and without them the brain is a changed place. The sleeper fails to waken, partly because the circuits that carry most incoming (sensory) and outgoing (motor) impulses are blocked at the brain stem, producing temporary anesthesia and paralysis. And while the sleeper is conscious (as indicated by dream memories and other data), the impulses churning through the brain are internally generated rather than external; the ways they are processed are also different; and so the liberated genie does not rouse the sleeper to deal with the outside world, but instead serves to evoke REM sleep and dreams.

The genie soon returns to its lamp (adults spend less than a quarter of their slumber in REM sleep), partly because the chemical excitement of the REM phase activates the neurons that distribute norepinephrine and serotonin, and that inhibits the acetylcholine generators. So norepinephrine and serotonin levels rise, excitement fades, and the sleeper progresses out of REM sleep and downward through Stages II, III, and IV, until the cycle is ready to repeat or the sleeper is ready to awaken.

Besides helping to explain the dynamics of REM sleep, this "reciprocal interaction" theory shone a powerful light on the mystery of dreams. Among other things, it indicated that REM dreams arise from an internal activation of the brain triggered by the brain stem and increasing acetylcholine levels. This broad activation elicits fragmentary information from whatever specialty shops happen to be activated, and the resulting hodgepodge gets stitched together by a brain skilled at recognizing both real and imaginary patterns. In other words, REM dreams do not spring whole from the subconscious and get cut by a censor to wind up resembling a patchwork. Rather, they start out as bits and pieces that get assembled and sewn into a patchwork.

As one might expect, emotion centers are active in REM sleep. Foremost among these is the amygdala, an almond-shaped structure deep in the cerebrum that generates strong emotions, most notably anxiety. The amygdala is selectively activated in REM sleep, which probably explains why strong emotions, especially anxiety, pervade REM dreams. At the same time, the dreamer's prefrontal lobes—housing the critical faculties that normally balance such emotions—lack the serotonin and norepinephrine they need to do their work and have been selectively turned off. So emotion has a field day.

Notice, for example, that the dream about the children has a sensual-erotic character throughout, while the dream about the restaurant is filled with anxiety throughout. This emotional consistency stands in sharp contrast to these dreams' fractured and bizarre plot lines—featuring such things as emerging dry from a full bathtub, having the pool turn into a beach, having an oil bag turn into a burlap bag, and not being burned by the hot oil. It is true that dreams can change their emotional flavor, but dream emotions generally appear far more consistent than dream plots; and it often seems (as when the waitress "dreads" having to start over) that the brain is trying to shape the plot to dream emotions rather than the reverse. Thus, for reasons relating to brain anatomy and chemistry, most REM dreams appear emotion-driven.

To an extent, the dreamer may be willing to accept odd plot lines because they help account for felt emotions. But the main reason why the dreamer accepts this zany stuff is that judgment—another function of the prefrontal lobes—isn't working. The dream registers in consciousness because the brain is very active, and because enough elements of consciousness are working. However, the shortfall in norepinephrine and serotonin (as well as a lack of stimulating brain stem signals to the prefrontal lobes) disables judgment. So the dream gets accepted uncritically, and is then mostly or entirely forgotten because this same chemical shortfall disables memory.

REM conditions also help to explain dream movements. The motor and visual parts of the cerebral cortex are very active in REM sleep, so dreams are very visual and often full of motion. But the motion is peculiar, because incoming sensory impulses are blocked at the brain stem, and so are outgoing motor signals. Thus, we may dream we are running but getting nowhere because the brain is issuing orders to run but we aren't moving. Or else, we may dream we are falling because our eyes are closed and the brain stem has blocked signals from our body that tell us the force of gravity has been thwarted, and so the brain (especially if it is registering anxiety and fear) decides that it is falling.

What all this shows is that we have a good handle on dreams. In essence, dreams appear to result from brain activation, and appear to be synthesized from smaller bits of information. This "activation–synthesis" theory has important implications for psychiatry, because it implies that dreams can be understood without resorting to symbolism, censorship, or other such devices. Indeed, they should be understood that way, because time and expense aside, interpretive approaches tend to get in the way by obscuring the dream's transparent meaning.

When Bob McCarley and I first presented this activation–synthesis theory in 1977, we were widely accused by dream interpreters and others of belittling the significance of dreams. But in fact we were not belittling dreams' significance. We were belittling the obscure gyrations of dream analysts because we believed those gyrations were not needed. We felt strongly (and still do) that dreams can give insight into mental problems. But such insight can be gained by psychologists, therapists, and counselors who are ordinarily sensitive, without any specialized training.

As this suggests, dreams can prove meaningful to both mental health professionals and dreamers who recognize certain basic facts. Among them: Dreamers tend to dream about things that concern them. Also, emotional feelings tend to drive dreams, and so they are filled with elements that reflect the dreamer's real-life emotional concerns. Beyond that, dream emotions and thoughts tend to get synthesized into transparently meaningful dream plots of therapeutic interest. But these require no high-flying interpreter, because they represent the brain's own best efforts to interpret the component elements. So therapists and patients should discuss dreams' transparent content, a process that besides being fruitful avoids the elaborate procedures and major drawbacks of Freudian and other "hidden meaning" methods.

This is not the end of what sleep and dream research implies for psychiatry. Rather, as pointed out in Allan Hobson's *Dream Drugstore*, it is only the beginning. For in some ways the chemical brain dance we have been watching resembles a masquerade ball, and we have reached a point in the proceedings where it seems right to have certain participants unmask. To wit: Dreaming is a form of psychosis. Our old friends the neuromodulators are prime agents of this and other forms of psychosis. Our psychoactive drugs, both medical and recreational, are in fact the handmaidens of these neuromodulators. And we can vastly improve our understanding of psychosis, our treatment of it, and our awareness of drug benefits and dangers by exploring the nightly minuet of sleep and dreams.

Let's start by defining psychosis. Psychosis is a pronounced inability to cope with reality, usually as a result of hallucinations or delusions or both, and is often accompanied by strong emotions—either because the brain is emotionally stimulated or because strong feelings are generated by frustration.

So how does dreaming measure up to our definition? Clearly, the conscious state of REM dreaming is psychotic. The dreamer cannot cope with reality. He can't even recognize reality. His critical judgment is on hold. His dream hallucinations are compelling. He is generally deluded into believing that all sorts of bizarre dream events should be accepted as reality. And the whole experience generally gets bound up with strong emotions. So the match is nearly perfect; and in fact the dream state seems about as psychotic as anything in the psychiatric realm.

But what causes this psychosis? We could take an anatomic approach and go on about how the visual cortex, the motor cortex, the parietal operculum, the amygdala, parts of the brain stem, and so forth are very active; how the prefrontal cortex and the memory centers in the hippocampus are inactive; how the thalamus and cerebral cortex are maintaining a conscious state; and how dreaming is the net result of all this. But that isn't really necessary, because something much more basic has happened that appears responsible for these regional effects. Namely, two major neuromodulators (serotonin and norepinephrine) have withdrawn from the dance, while two others (acetylcholine and dopamine) hold the floor.

If serotonin and norepinephrine return in force, the result is quick and dramatic. The regional brain effects change, the sleeper wakes, dream psychosis ends, and the brain passes into its normal waking state. Of course, the dreaming brain is working with internal information lit up by signals from the brain stem, rather than with external information received from the real world, and this contributes to the psychotic state. But if one is seeking a prime mover, it seems fair to say that the observed neuromodulator imbalance (serotonin and norepinephrine down, acetylcholine and dopamine up) is the main cause of dream psychosis.

Of course, we shouldn't pretend that all psychoses are alike, because in fact they differ greatly. Psychoses associated with schizophrenia commonly involve accusing voices (auditory rather than visual hallucinations), paranoia (rare in dreams), and a flat emotional tone. About the only feature shared with dream psychosis is some not very specific anxiety.

Or consider another variation. Psychotic states tied to depression are typically obsessed with disease, death, and rotting or missing body parts—possibly because they involve the emotions of sadness, shame, and guilt that are generally absent from dream psychoses.

The psychoses of mania do share some features with their dream-related cousins, because dream psychoses can have the delusional grandiosity and emotional elation found in mania. However, as in schizophrenia, the manic psychoses lean toward paranoia and toward auditory rather than visual hallucinations.

The psychoses that seem most dreamlike are those of toxic delirium caused by psychoactive drugs like LSD and also by alcohol abuse. Even here there are various differences. Most notably, toxic delirium generally occurs while the subject is awake. But as in dreaming, the hallucinations tend to be visual, and memory problems commonly produce disorientation and amnesia. So these psychoses seem inclined to share something with the dream state that the other psychoses lack—a likely possibility being strong deactivation of serotonin.

On a larger canvas, it may be more productive to ask what the psychoses have in common than to focus on how they differ. That's because we know what happens in REM sleep. The brain activates regions that produce emotions and internally generated imagery, and these work together to create psychotic dream plots. At the same time, the brain blocks external sensory inputs (needed for orientation) and selectively deactivates parts of the prefrontal cortex (thus disabling self-reflective awareness, insight, judgment, and executive guidance).

Surprisingly, *all* psychoses seem to share these or similar underlying elements. That is, they all seem to tilt the brain's balance in favor of emotions and internal perceptions that work together against thought. So it seems likely that all psychoses arise from a changed brain activation pattern that shifts this balance. Experience with REM sleep shows that one way to produce this shift is to disable the serotonin and norepinephrine systems, allowing the acetylcholine and dopamine systems to run free. This produces all of the various brain activation changes needed to create psychosis; and conversely, restoring serotonin and norepinephrine to waking levels ends the psychotic state.

If the generalized chemical pattern is serotonin/norepinephrine down, acetylcholine/dopamine up, it seems reasonable to ask whether all the psychosis generators perform this minuet. They all seem to, but they all use slightly different dance formations. Schizophrenia raises sensitivity to dopamine; mania raises acetylcholine; depression lowers serotonin; and REM sleep sweeps serotonin and norepinephrine off the floor.

Of course, dancers don't work in isolation and neither do neuromodulators. So giving a greater or lesser role to one is very likely to affect the others, and the end result is not the product of one but the sophisticated interaction of them all. What's more, these are not the only dancers present, because the brain has other modulators; and besides the specific formations chosen, other factors like how active the dance is (the level of brain activation) and the use of natural or artificial scenery (exterior or stored imagery) strongly influence the scene.

Even so, what we are suggesting here is reinforced by what we know of the psychiatric drugs. For if all psychoses arise from similar formations in this chemical dance, then reversing those formations should reduce psychosis. And in one way or another, that is precisely what all the leading psychiatric drugs actually do. Some, like the older neuroleptics (e.g., Thorazine), reduce the effects of dopamine and (to a lesser extent) acetylcholine. Others, like the monoamine oxidase inhibitors (MAOIs), enhance the effects of serotonin and norepinephrine. Still others, such as the newer "atypical" antipsychotics (e.g., Clozaril), enhance serotonin while counteracting dopamine. But regardless of the modulators they affect, these drugs and others (including the tricyclic antidepressants and the selective serotonin reuptake inhibitors like Prozac) all work to enhance the effects of serotonin/norepinephrine, reduce the effects of dopamine/acetylcholine, or do both things at once. This strongly supports the thesis that while the various psychoses have different medical and anatomic causes, as well as different symptoms, a common process underlies them all—the same recurring chemical process that creates our dance of dreams.

7

Mapping Inner Space: New Models of Mental Order and Disorder

Not all patients on Prozac respond this way [highly favorably]. Some are unaffected by the medicine; some merely recover from depression, as they might on any antidepressant. But a few, a substantial minority, are transformed. Like Garrison Keillor's marvelous Powdermilk biscuits, Prozac gives these patients the courage to do what needs to be done.

—**Peter D. Kramer,** *Listening to Prozac,* 1993

Natural science tends to be precise; but sometimes our observations are less precise than one might think. Jonathan Leonard recalls the day this point was driven home at a Harvard chemistry lecture he attended as an undergraduate. The professor, a fellow named Nash, took out an elegant model of an atom and put it on the table before him. Like most classical atomic models, it showed the atom's electrons circling the nucleus in precisely defined orbits. "But it's not really like that," Professor Nash explained. "It's really more like this!" And from under the table he took a large ball of glass wool, which he placed on the table for comparison beside his model.

As our dancing neuromodulators suggest, the precise nature of mental ills can be just as hard to pin down as the precise location of electrons. That has major implications for diagnosis and treatment. So before getting into specific ailments (anxiety disorders, depression, and schizophrenia), we are going to pause briefly to look at this question of precision and the general nature of mental ills.

What *can* brain science tell us about this? For starters, it can show why mental ills are both persistent and changeable. They are persistent because they arise mostly from structural problems or hard-wired memories that defy easy alteration. But things don't stop there. In many cases, as will be seen, these problems alter neuromodulator balances; or else the brain tries to adapt to them with countermeasures that alter those balances. This means that many symptoms arise from such imbalances. And since the brain can swing neuromodulator balances enough to make each of us psychotic several times a night, it's hardly surprising that many overt symptoms of a given mental illness can change too.

We can get a better look at this with a model that one of us (Allan Hobson) devised to help explain things to mental patients. The model, which focuses on brain chemistry, shows how the brain shifts smoothly from one state to another. The model's main drawback and also its greatest strength is its simplicity. That's because it considers only three variables and ignores regional and structural differences within the brain. This may seem unwise to many neurologists, who deal mainly with local problems (strokes, tumors, etc.), and it could encourage us to overlook important regional variations in visualizing mental ills. But the reverse is also true; for it seems likely that many symptoms of mental illness arise from chemical imbalances induced by brain coordination or other problems that involve many different structures. And in these cases we might well be better off focusing on this model and the changes that it registers rather than on specific local variations.

The model's three variables are the brain's **A**ctivation level, **I**nformation source, and **M**odulation pattern. This so-called "AIM" model (Figure 7.1) takes the form of a three-dimensional cube, with one dimension being devoted to each of the three variables. Looking first at the cube's front, we see that **A**, the brain's activation level (represented by EEG activity) starts at a

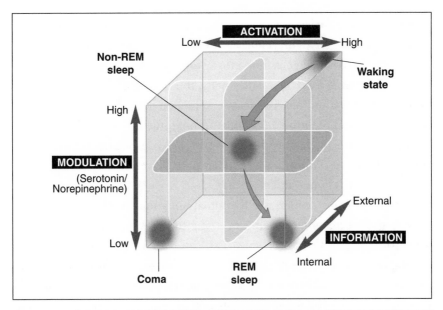

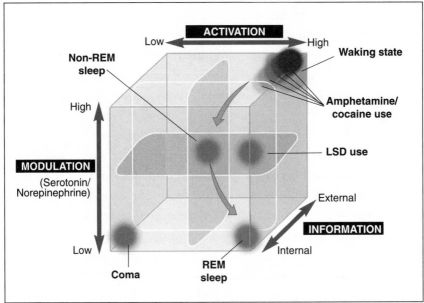

FIGURE 7.1   Depictions of the three-dimensional AIM cube charting various brain states in terms of EEG activation (A), use of external or internal information (I), and serotonin/norepinephrine modulation (M). The cubic part A, top, shows the normal waking state, as well as the states of non-REM sleep, REM sleep, and coma. The cubic part B, bottom, shows additional states induced by LSD and amphetamine/cocaine use.

minimum on the far left and increases (with increasing consciousness) to a maximum on the far right. Similarly, the nature of sensory information (**I**) that is being processed goes from mostly internal at the front of the cube to mostly external at the back. And finally, the degree of serotonin/norepinephrine modulation (**M**) rises from a minimum at the bottom of the cube to a maximum at the top. Serotonin and norepinephrine are the only two neuromodulators tracked by this model, because despite occasional exceptions a high serotonin/norepinephrine level tends to work against dopamine/acetylcholine and vice versa.

Clearly, the reality underlying this situation is so complex that we admit to feeling a bit like military officers trying to chart the ebb and flow of a great battle with little flags. Even so, we can use the AIM model to locate just about any brain state in the virtual space created by our three very real dimensions. For instance, where is a patient in a coma? In a coma, brain activity is minimal, production and distribution of serotonin/norepinephrine is likewise minimal, and the brain is using internally generated information. So the comatose brain is at the left lower front corner of the cube. Similarly, a normally awake brain is at the opposite (right upper back) corner of the cube—because brain activity is high, levels of serotonin/norepinephrine are also high, and the brain is working mainly with external information.

What happens when you go to sleep? If you go to bed because you feel tired, your levels of serotonin/norepinephrine are falling, so you are already moving lower in the cube. Then you shut your door (to reduce sound), lie under blankets (muffling touch sensations), and close your eyes, thereby moving from the back of the cube toward the front. Gradually, as you enter progressively deeper stages of non-REM sleep, your brain does less work— as indicated by the synchronous discharges that show up as "slow waves" on the EEG—and so its activity level moves from right to left.

You are now in the land of Stage IV sleep. At this point the brain stem gets its cue from the hypothalamus, the locus coeruleus and raphe nucleus shut down, and serotonin/norepinephrine levels plummet, pulling you to the bottom of the cube. But then acetylcholine levels (which had been falling as you drifted into Stage IV) shoot upward; the brain shifts into high gear (moving you to the far right-hand side of the cube); and you find yourself conscious but in dreamland, working with internally generated data (at the front face of the cube). So you are now at the bottom right front corner of the cube in REM sleep.

As we have seen, REM sleep is psychotic. But it's not surprising that this lower right corner of the cube should signify a psychotic state; for the brain is very active, but the virtual absence of serotonin and norepinephrine disable thought and judgment.

This is somewhat like the psychotic state produced by LSD (see Figure 7.1, bottom). Here again the brain is very active. And here again the subject moves toward the bottom of the cube—because LSD enhances the dopamine system while lowering serotonin levels. The enhanced dopamine system probably heightens the LSD tripper's state of arousal; but the reduced serotonin tilts the brain toward internal imagery and emotions, and away from judgment and reason. As a result, the tripper travels from the back of the cube toward the front, far enough forward for internally oriented sensory processing to bend the truth and create visual hallucinations, but not so far as to abolish all external information. The result is what has been called a "waking dream."

Such visual hallucinations also feature prominently in psychotic states arising from alcohol, opium, and heroin addiction. In contrast, psychoses associated with amphetamine and cocaine addiction (and also with schizophrenia and mania) do not typically cause visual hallucinations but instead rouse imaginary voices. How might we explain this difference?

We know that the amphetamines and cocaine tend to enhance the dopamine system and alter other neuromodulator levels in ways that promote unbridled stimulation—producing feelings of unclouded consciousness, enhanced mental energy, elevated mood, increased physical strength, and great sexual potency. In sum, we find that cocaine and the amphetamines can push us through the AIM model's roof. That is, these stimulants appear to greatly enhance the waking state. So while the brain is fully active, highly modulated, and wide awake (in the upper right-hand part of the cube), the waking state has in fact been extended beyond its normal limits. So the starting user may gain superior abilities and may reach a point *above* the back right corner of the cube. But in this unnatural state the hyperactive brain may also fail to coordinate rationally and may focus on internal signal processing to an extent that can produce psychosis. In this latter case, the user is still at the top right of the cube but has moved from the back of the cube toward the front.

This suggests that there may be two principal kinds of psychosis. In one of these (the REM sleep, LSD, alcohol, and opiate type), the thought and memory systems located toward the front of the brain are so disabled that they cannot make sense of pseudosensory signals processed by the back of the brain, and so they report credible visual hallucinations. In the other kind of psychosis (the cocaine, amphetamine, schizophrenia, and mania type), the forebrain's thought and memory systems get cranked so high that the brain becomes hyperattentive, fails to coordinate properly, hears accusatory or threatening voices, and is prone to paranoia. This suggests that sanity depends on a delicate modulatory balance generating neither too much forebrain activity nor too little. And while we normally operate within safe limits, we find that various drugs and mental ills can push the system too far in one direction or the other.

Besides encouraging such insights, the AIM model is useful in the clinic. We all know that mental patients have a tough time adjusting to their plight. They often feel plagued with an affliction that has boxed them in, stigmatized them, and set them apart from others. The AIM model can help them to understand their illness in terms of things the brain does normally. That is, it can tell them about the nature of their illness while demonstrating that they have a lot in common with normal people—all of whom go through several psychotic REM sleep episodes every night. So it can reduce their sense of stigma, encourage understanding, and foster hope.

But the AIM model has far broader implications. We can see this if we compare its simple picture of brain state continuity to the large and confusing picture enshrined in the American Psychiatry Association's 1994 (fourth) edition of *The Diagnostic and Statistical Manual of Mental Disorders* (DSM-IV). This classification manual is anything but simple. In fact it is so complex that it lists over 1800 distinct diagnostic mental conditions. And since it is the profession's official diagnostic guide, it gets lots of use and tends to focus attention on small differences.

One of us (Allan Hobson) recalls a day some years back when he heard Lewis Judd, then Director of the National Institute of Mental Health, tell the Dalai Lama that there were 1800 discrete diagnostic conditions defining mental illness. Frankly, that's not possible. It makes no sense. The categories are too finely divided. If you are in category 1750 today you will just as likely be in category 1756 tomorrow, because your brain has slipped across some arbitrary boundary set out in the manual. That makes these boundaries suspect, because some of them are real while others simply reflect more or less mercurial brain states.

This official classification system makes it tempting to pigeonhole patients and prescribe psychiatric drugs by rote: "Let's see now, if Mrs. Smith has condition XYZ she should get drug 25 at 10 mg/day." Of course, most experienced psychiatrists realize that mental ills defy this sort of pigeonholing and respond poorly to such cavalier treatment. Even so, DSM-IV's authoritative status and detailed nature tends to promote the idea that rote diagnosis and pill-pushing are acceptable.

In striking contrast, the AIM model views all mental conditions as part of a single brain-state continuum. So it exposes the shortcomings of DSM-IV's hair-splitting approach. It also highlights the variability of mental ills, points up the need for monitoring and treatment continuity, shows why drug prescription needs to be based on understanding of the patient, and rightly debunks the notion that an isolated prescription emerging from some fifteen-minute quick fix appointment should be clothed even marginally in the raiment of sound treatment.

Of course, this infinite variability of brain states is only one thing that makes psychopharmacology tricky. Another is the nature of the drugs themselves. There are well over a hundred different psychiatric drugs. They have complex effects, need different lengths of time to become effective, and can influence different sorts of mental states in different ways. All these details need to be learned carefully. There is reason to suspect that psychiatric residents don't always succeed in this endeavor, even when they speak perfect English—much less when they come from a foreign land and their command of English is uncertain. This may be one reason why less than half the foreign residents succeed in passing the board examinations

they must pass to become practicing psychiatrists. In any event, these circumstances argue strongly for closely linking pharmacology to therapy and
emphasizing treatment continuity—because otherwise a case may slide
along without anyone realizing that the patient's mental state has changed
and a new drug may be needed, or even that the patient has been getting the
wrong drug all along.

That's only one part of the drug problem. Another is side effects. These
range from minor to debilitating or even fatal. For instance, one class of
older antidepressants, the MAOIs (monoamine oxidase inhibitors), can induce life-threatening hypertension if the patient consumes any of various
foods or other medicines on a warning list that includes cheese and cough
syrup. One of the newer antipsychotics, clozapine, can sometimes cause a
deadly blood disorder called agranulocytosis. And many antipsychotics,
most notably the classic ones like Thorazine that block the D_2 dopamine
receptor, can produce a Parkinsonian syndrome called tardive dyskinesia.

Tardive means delayed, and this syndrome (consisting of involuntary
movements, especially facial movements, and rigidity) can take more than a
decade to emerge. Unlike more manageable Parkinsonian symptoms, which
emerge soon after initial use of an antipsychotic drug, tardive dyskinesia is
hard to treat and may prove permanent. So it is rightly feared as an extremely embarrassing, debilitating, and undesirable side effect.

On the positive side, we now have good alternatives to the MAOIs, and
many of our newer antipsychotics seem less inclined to cause tardive dyskinesia. That could be because some of them have not been around long
enough to get evaluated fully. But many of these newer drugs could also be
causing fewer Parkinsonian effects and less tardive dyskinesia because they
are targeting neuron ports other than the dopamine D_2 receptor.

Be this as it may, beyond such devastating problems lies a wide array of
lesser side effects such as muscle tics and twitches, reduced sex drive,
weight gain, sleepiness, and excess salivation. Though these are certainly
lesser problems, they cannot be ignored, for they are common and can
cause major inconvenience—plenty in some cases to prompt the patient to
abandon treatment. So here again we can see that drug prescription should
not be an isolated event based on rote diagnosis, but instead should be

closely tied to thorough understanding of the patient's case, as well as to monitoring and therapy.

We are unlikely to escape this motley collection of significant side effects anytime soon, because virtually all of our psychiatric drugs affect one or another of the brain's neurotransmitter or neuromodulator systems. This can dispel or reduce a wide range of mental symptoms, depending on the mental disorder being treated and the drug used. But that is only a small part of what is happening. For in fact the brain's chemical state is being changed—moving the patient into partially uncharted or even forbidden regions of the AIM state space. So besides having beneficial effects, the drug is likely to cause other effects that are neutral or adverse. And besides making immediate or short-term changes, the drug is likely to slowly alter how neurons respond to the new chronic excess or shortfall of one or more affected neuromodulators—thereby changing the brain in ways that are unanticipated, often hard to reverse, and sometimes permanent.

This raises the larger issue of how psychoactive drugs are being managed—not just by psychiatrists but by general practitioners and average citizens. Recalling the bygone days when Coca-Cola contained cocaine and morphine could be bought over the counter, we can see a persistent human tendency to use powerful psychoactive drugs for casual or recreational purposes. That trend wasn't stopped by making certain drugs illegal. And looking beyond the traffic in illegal drugs, over the past half century we can see this same inclination to use strong drugs for less than compelling reasons being applied to certain psychiatric medications. As we know, the "tranquilizer" Miltown created a sensation before Valium and other benzodiazepines came along; then Valium held center stage as the world slowly acknowledged that both Miltown and the benzodiazepines were addictive; and now public attention has shifted to Prozac, Zoloft, and other SSRIs (selective serotonin reuptake inhibitors).

Overall, the prevailing pattern is one of massive ill-regulated use. Naturally, this problem is not limited to the SSRIs; but SSRIs are all the rage right now, and they provide a suitable example. We aren't sure exactly how many Americans are now taking SSRIs, but rough estimates suggest that over 5% of all adults have used Prozac at one time or another. Beyond that, a good many more people are on other prescription SSRIs like Zoloft and Paxil; and additional multitudes are taking the "dietary supplement" St. John's wort (estimated sales of US$400 million in 1998), which appears to act as an SSRI.

Even the prescription SSRIs are commonly prescribed by general practitioners rather than psychiatrists; and since St. John's wort is an unregulated dietary supplement, it can be purchased with no prescription. The advantage of such informality centers on the fact that these drugs are commonly used to fight depression. Obviously, depression can be a dangerous condition that can lead to suicide. But many people with depression, even severe depression, fail to seek out mental health professionals—partly because of the stigma associated with mental illness, and partly because they are depressed and therefore not highly motivated to seek care. So one might argue that current arrangements are pretty satisfactory.

But the low proportion of depressed people treated for depression and current suicide statistics suggest otherwise—partly because general practitioners often fail to spot depression or fail to treat it, and partly because depressed people have a poor understanding of how to treat themselves. What's more, history shows that self-medication with strong drugs is rarely the right answer. If we really want to fight depression, then depression should be destigmatized as much as possible; people should be better informed about it; depressed people should be given greater incentives to seek care; and those who do seek care should be referred to a center that can provide not just a prescription but appropriate therapy, community services, and follow-up.

None of this is any panacea. But neither is St. John's wort, most of whose active ingredients are not standardized and whose mechanism of action is unclear. Though a 1996 review of the literature, published in the *British Medical Journal,* found St. John's wort to have antidepressive effects in cases of mild to moderate depression, no long-term studies have been completed and the optimum safe dosages of this herbal remedy are unknown.

If such ignorance gives grounds for caution, so does our emerging knowledge of Prozac and the other regulated SSRIs. To begin with, they make a frontal assault on the architecture of normal sleep—prompting

vivid dreams and sometimes dream-related movements, drastically altering the nature of Stage I sleep, and throttling back on the other sleep
stages. If we home in on Prozac, Hugo Rosen's *Clinical Psychopharmacology for the Busy Practitioner* (second edition) cites a long list of possible
Prozac side effects, including headache, nervousness, insomnia, tremor,
anxiety, fatigue, nausea, vomiting, diarrhea, loss of appetite, and constipa-
tion. Also, according to Harvard Medical School psychiatrist Joseph
Glenmullen, author of *Prozac Backlash*, sudden exposure to Prozac has allegedly prompted a limited number of homicides and suicides; and a large
share (roughly 60 percent) of all Prozac users are said to experience sexual
dysfunction—including decreased libido, decreased arousal, and delayed
orgasm.

These side effects show more or less what we would expect from one
leading member of a powerful class of psychoactive drugs that encourage a
major neuromodulator to alter the brain's state. So these effects are cause
for concern. They are the kinds of effects that prompt patients to abandon
treatment. They give reason for close monitoring and communication with
the patient. And they raise a red flag against casual use.

But casual use is what we're getting. Besides being utterly unregulated,
St. John's wort is commonly used not only against depression but, more
imaginatively, against AIDS and a broad range of neurologic ills. And
while some people take Prozac and other SSRIs to treat depression or
anxiety disorders, others use them to lose weight, prevent premature
ejaculation, fight premenstrual syndrome, or reduce back pain. And still
others use them because it makes them feel sharp and improves their job
performance.

Unfortunately, this is only part of a far more general pattern involving the
amphetamines, the benzodiazepines, other widely used drugs like Ritalin,
and a number of other psychiatric medications. It is the kind of pattern we
would expect psychiatry's drift into depersonalized biomedicine to promote; but it is a pattern fraught with problems, one that seems clearly contrary to the public interest. This is something that goes beyond the variable
nature of mental ills or the need to effectively coordinate medication with
therapy and follow-up. For clearly, the mental health field should not be en-

gaging in quick fixes that may not work and that tend to perpetuate this pattern. Rather, it should recognize its moral obligation and should try to set an example, not only by prescribing drugs carefully but by keeping people informed about the true nature of psychiatric medications at a time when that nature is increasingly being illuminated by our growing understanding of brain states and brain chemistry.

All this reminds us of a story that took place nearly half a century ago. In the depths of the Cold War, sometime around the 1950s, those who felt the United States ought to do more nuclear testing had to justify that stance. So they started pushing an odd weapon called the "neutron bomb" that had never been developed and would require bomb tests in order to get built.

The neutron bomb was not some super-destroyer. Instead it was billed as a "tactical" nuclear device. Theoretically, the nuclear fires within in it would never burn full-force. Designed to release very little of its devastating energy, it would instead emit an invisible burst of neutrons. The neutrons would kill people out for a mile or so, but would leave buildings and most other goods intact with little or no residual radioactivity. So in theory the invaders could come in and seize this real estate for the mere price of removing their enemies' dead bodies.

Whether or not this theory could be made to work in practice, it seems clear that it had a fatal flaw. Namely, nothing could be done to keep the neutron bomb from being a nuclear device. So using it in a nonnuclear war would up the stakes and could conceivably unleash a nuclear Armageddon. Even if this chance were small, it would hardly be worth the price of the bricks and mortar saved. And of course, if a real nuclear war had already broken out, occupying a few enemy cities in a way that saved some real estate would hardly be a priority concern. Therefore, even if this bomb could be built, its benefits would fail to match the significant atomic risk.

It turns out that deploying the neutron bomb is a bit like using psychoactive drugs for casual purposes. The stakes are just too high; indeed, it seems hardly more advisable to treat minor ailments with psychiatric drugs than it is to shoot pheasants with elephant guns. For the drugs, like their recreational counterparts, alter basic brain chemistry. That can do good things and bad. The trouble is, unless we have long experience with a particular drug, we don't know where the balance lies. What's more, every person is slightly different; the brain's long-term responses can differ greatly from its short-term responses; and so, in cases where a brain drug isn't really needed or is only marginally needed, using it amounts to playing with psychiatric fire.

That doesn't mean we should turn away from psychiatric drugs. We all know they can dispel symptoms for people who would otherwise be crippled by mental ills. But we should prescribe drugs carefully; we should properly monitor the drugs that we do prescribe; we should publicly discourage large numbers of our fellow citizens from using these drugs to serve ends that seem merely fashionable, cosmetic, or convenient; we should note the neutron bomb problem; and we should recall the folly of using an elephant gun to shoot a pheasant.

PART THREE

Psychiatry and the Brain

8

Anxiety and
the Fear Machine

Charles Darwin (1809–1882) suffered from a chronic illness
that, throughout much of his adult life, impaired his
functioning and severely limited his activities. The writings
of this famous scientist as well as biographical materials
indicate that he probably suffered from an anxiety disorder.
His symptoms, when considered individually, suggest a
variety of conditions, but taken together they point toward
panic disorder with agoraphobia. This diagnosis brings
coherence to Darwin's activities and explains his secluded
lifestyle, including difficulty in speaking before groups and
meeting with colleagues.

—**Thomas J. Barloon and Russell Noyes, Jr.,**
Journal of the American Medical Association,
277: 138, 1997

So far we have been dealing with broad subjects—consciousness, brain states, and the general nature of mental ills and psychiatric drugs. That's fine as far as it goes, but it's not very specific. To get specific, we need to examine what brain science has discovered about particular kinds of mental ills. Therefore, in this and the next two chapters we will focus on three of psychiatry's main concerns: the anxiety disorders, depression, and schizophrenia. Each of these has been the subject of many large tomes and vast clouds of scientific papers, so we don't expect to cover the whole territory. Instead we will describe what we find interesting; we will examine certain discoveries; and we will show that brain science has reached a point where it can provide a sound theoretical and practical foundation for treating these diseases of the mind.

I had no childhood concern about pheasants or elephant guns, Jonathan Leonard recalls, but I did hate pig piles. Most days would be all right. But every so often, on the school playground or in some kid's backyard, a bunch of playmates would decide to pile on top of one another, and I would wind up somewhere near the bottom. That seemed fine with everybody else, but it wasn't fine with me. I couldn't move. I couldn't breathe. I was suffocating, panicky, desperate. Half of me, the smaller rational half, said not to worry. The larger terrified half said I was smothering—a fate worse in my hyped-up imagination than any other death. So if the pileup ended fast, well and good. If not, I would yell "Help!" from the bottom of the pile until the other kids got off with puzzled looks and asked what could possibly be wrong. Trying to explain that I couldn't breathe was a bit embarrassing, but embarrassment was nothing compared to the huge unbridled fear I felt at the bottom of the pile—fear caused by being a little claustrophobic.

Of course, I was not alone. Lots of people have claustrophobia (fear of closed-in spaces) at least as severe as mine, and lots more have other variants like animal phobias (fear of dogs, snakes, bugs, spiders, mice, etc.), acrophobia (fear of heights), agoraphobia (fear of situations from which escape may be difficult or embarrassing), or any of various social phobias (fear of embarrassment in social situations). In fact, if you add them all together, phobias (excessive fear of specific situations or things) are about the commonest of all mental problems. It's hard to tell just how common because it's hard to set standards (phobias range all the way from barely perceptible to crippling), but the U.S. Surgeon General's 1999 report on mental health suggests that roughly 8 percent of all American adults have some sort of "simple" phobia (simple phobias do not include agoraphobia or social phobias), this being about half the total share of Americans (16 percent) reported to have some kind of diagnosable anxiety disorder.

Thus, claustrophobia is only one kind of phobia, and the phobias are only one kind of anxiety disorder. What's more, as their name implies, the anxiety disorders involve not just inappropriate fear but an array of inappropriate feelings that can range from mildly excessive nervousness to overwhelming terror. What identifies these disorders is the abnormal nature of such feelings (anything from chronic worry about minor or unlikely problems to panic for no obvious reason). That's because the whole spectrum of feelings that includes nervousness, concern, anxiety, fear, and terror is perfectly normal, often helpful in directing attention, and sometimes vital to survival. It is only when such feelings are not normal (when they are out of proportion to the problems at hand, have no obvious cause, or persist well beyond the time when they are useful) that they point to a possible anxiety disorder.

Of all the anxiety disorders, the phobias are among the most commonly ignored. According the National Institute of Mental Health, less than a quarter of all people with phobias ever have them treated. That's probably because most of us who aren't badly afflicted feel we can deal with it— we've lived with it most of our lives, after all, and it seems a minor matter. Also it seems so well entrenched, so much a part of us, that the strong measures needed to beat it down with therapy or drugs seem likely to be more trouble than they're worth.

But that doesn't stop us from being curious; nor does it dispel the notion that learning about phobias could teach us important things about the brain. So it seems reasonable to ask: What causes phobias? Why do different people have different ones? Why do they jerk us into emotional overdrive? If we can't escape from the triggering situation, why are we seized by irrational all-consuming panic? Can anything be done about it? And why do our fundamentally rational but survival-oriented brains bring us to this sorry state?

Survival, of course, is the key. Our brains are trying to deal with life-threatening perils like being trapped, getting crushed, falling, being bitten or attacked by dangerous creatures, getting lost, or (in anthropological terms) being rejected (and so perhaps killed or ejected) by the tribe. These are not minor matters. They are all-or-nothing issues. So the brain has built in a set of all-or-nothing responses; and since many threatening situations need quick answers, it has set these responses to go off automatically when triggered by events thought to be associated with some peril.

The process of deciding which events will act as triggers is called conditioning. We have known about conditioned responses for a long time, ever since the start of the twentieth century when Ivan Pavlov trained hungry dogs to salivate at the sound of a bell. Of course, many sorts of responses, not just life-and-death ones, involve conditioning. But since phobias and certain other anxiety disorders seem strongly tied to the conditioning process, it is worth looking at what conditioning is and how it works.

In its pure and classic form, the conditioning process involves an "unconditioned" and a "conditioned" stimulus. For example, if food (serving as the unconditioned stimulus in this case) is placed within the sight of a hungry dog, it will cause the dog to salivate. By repeatedly ringing a bell (Pavlov's conditioned stimulus) to signal arriving food, the dog can be trained to salivate at the sound of the bell whether or not food actually arrives.

A process called fear conditioning works much the same way. We can fear-condition a rat by sounding a tone shortly before giving the rat a mild shock through the mesh grid on the floor of its cage. After some repetition of these steps, the rat will show marked anxiety after the tone, even if it receives no shock. Among other things it will stop moving (a "freeze" response to minimize chances of detection), its heartbeat will speed up, its pain sense will be suppressed, and its endocrine system will release stress hormones. It can also be trained to jump over a small barrier to avoid the shock, and will continue to jump the barrier at the tone even if no charge is delivered to the grid.

In a similar vein, suppose a monkey with an innate fear of leopards has a near-fatal encounter with a leopard at a watering hole. As a result, the monkey becomes wary around this and other watering holes. That's sensible and promotes survival. But we can have too much of a good thing. It is possible to imagine the monkey being so badly scared that it develops an extreme watering hole aversion, to a point where thirst becomes life-threatening. In that case the monkey now has a phobia—a watering hole wariness so excessive that it constitutes a phobic disorder.

Roughly the same process applies to people. We inherit a strong drive to keep breathing. So if we find that being closed in (under a pile of kids or whatever) interferes seriously with breathing in one or more cases, we may develop a generalized fear of closed-in spaces (claustrophobia). Or if some inherited aversion to snakelike things is reinforced by our own youthful experience with snakes or by seeing our parents give a fearful response to snakes, then we may develop a phobic reaction to snakes that applies not just to poisonous serpents but to any snakelike things (eels, salamanders, night crawlers), regardless of whether they are caged, even when we know perfectly well that there are no dangerous serpents for a hundred miles.

As this suggests, evolution tends to err on the side of safety. It generally doesn't hurt to be afraid of harmless snakelike things, but failing to fear a water moccasin (or even taking the time to identify it) can be fatal. So your brain tends to generalize. If it looks like a snake, don't stop, don't think, GET OUT. And if this sounds like a panicky overanxious response to a garter snake in your backyard, that's exactly what it is—with a purpose.

Of course, poisonous snakes have been around a long time. Evolution has had maybe a hundred million years to work with the ancestors of humanity and to favor the survival of those who avoided snakes. So we probably carry some inherited neural patterns that need only gentle prodding (like viewing others' aversions) to generate a lifelong emotional fear of these creatures. But the same does not hold true for modern devices like cars, chain saws, and handguns. So our brains are not primed to fear them, which probably goes a long way toward explaining why snake phobias are common while phobic responses to much more dangerous modern devices are rather rare.

This information, much of it known for a half century or more, gives a fair idea of what is happening, but it's rather fuzzy. To go further, to show how mental processes can account for specific features of phobias and other anxiety disorders, one really needs to look inside the brain. Fortunately, thanks to recent scientific advances, we can do precisely that.

It turns out that the headquarters of fear and anxiety in the brain is a structure called the amygdala. The amygdala actually consists of two small, almond-shaped bodies (one in each cerebral hemisphere) embedded deep in the brain close to the thalamus. As previously noted, the thalamus acts as the first checkpoint for virtually all incoming sensory information, which it relays to other parts of the brain. One place it relays such information is the amygdala, with which it has strong ties. We know that cutting these thalamus–amygdala ties prevents fear conditioning; and we also know that one collection of gray matter within the amygdala (the central nucleus) has the power to order up virtually every brain and body response associated with conditioned fear. This suggests that the thalamus is sending raw (essentially unprocessed) sensory information to the amygdala; the amygdala is matching this up with patterns in its own neural circuits; and if the match is close enough, it is unleashing the vast power of the conditioned fear response without waiting for guidance from above.

Jonathan Leonard's father, who had a fine sense of mischief in his youth, once carved and painted a snake-shaped fallen limb to look like a big serpent. Then in summer he would put it out behind some backyard bush at his Cape Cod home, along a path where friends and relatives would stroll. The number of people frightened by that snake was never tallied, but it was high—plenty high enough to show informally that the victims' brains had moved into high gear long before they got their bearings and realized that this serpent was a fake.

But of course, people did get their bearings. That's because visual information about the snake (a more refined version of what the amygdala got) was sent to the visual processing centers of the cerebral cortex, was processed there, and was used to generate messages able to refine or countermand the fear signals coming out of the amygdala. For this latter purpose, the brain has neural pathways running from sensory processing areas of the cerebral cortex to the amygdala. These sensory cortex–amygdala pathways go to the same part of the amygdala (the lateral nucleus) as the shorter, quicker pathways coming in from the thalamus. Thus the lateral nucleus is in a good position to coordinate these two data flows and relay them to the amygdala's command center (the central nucleus) so that the central nucleus knows whether to keep sending fear signals to other parts of the brain and keep turning up the body's defensive reactions or whether to quash its initial alarm signals—if in fact the purported snake is identified as a stick or for some other reason the imagined danger is found to be unreal.

But there's more to it than that. Memory plays an important role in what we consider dangerous. Depending on our own personal experiences and what others have told us, we will remember specific places as being dangerous—an unfenced road near a cliff, some traffic intersection, a bad part of town, a "haunted" house. Of course, identifying these places depends on more than sensory processing. It depends upon many pieces of coordinated information being matched up collectively with something we associate with trouble. Obviously, making such a match depends on memory.

Brain scientists commonly talk about two basic kinds of memory. One kind, called "procedural" or "implicit" memory, tells us unconsciously how to walk, talk, play a piano, drive a car, serve a tennis ball, and so on. Such memory, typically residing wholly or partly in places like the cerebellum and brain stem, does not much concern us here. The other kind of memory, known as "declarative" or "explicit" memory, allows us to recall all sorts of things—where we went yesterday, the sound of a parent's voice, how our car looks, or what we must know to pass a test. This kind of memory does concern us here, because it allows us to compare our current situation with all sorts of anxiety-generating circumstances, and so it is intimately connected with the amygdala's work and with anxiety.

Interestingly, the heart of our explicit memory system lies not in the cerebral cortex but in a pair of curved structures (one in each cerebral hemisphere) hunkered down deep in the cerebrum near the thalamus and amygdala. Known jointly as the hippocampus ("sea horse" in Latin), they play a vital role in explicit memory processing. We know this because damage to the hippocampus interferes with memory. Indeed, it interferes with memory so much that one famous patient, a man with the initials "H.M.," entered a time warp worthy of the *Twilight Zone* when his two hippocampal structures were removed surgically, together with portions of his temporal lobes, in an effort to quell his epilepsy.

The epilepsy was quelled, all right; but H.M. lost his ability to create new memories. He could recall old memories—things that happened before the operation—just fine. But he could only hold new memories of the explicit sort in his head for a few minutes. So if you met him, he might seem fairly normal. But if you left the room and returned a few minutes later, he would not know who you were. And even though he could learn new skills and procedures (because implicit memory does not depend on the hippocampus), his conscious mind was so completely trapped in this time warp that he could not identify himself in current pictures, but only in ones taken around the time of the operation and before.

This and other evidence singles out the hippocampus as the key player in formation of explicit memory. But of course the term "explicit memory" covers a lot of ground. Speaking broadly, virtually all the connections in the brain can be defined as "memory" of one sort or another; and much of the brain, especially the cerebral cortex, deals with matters relating to explicit memory. So how does the relatively small hippocampus get to play such a big role?

What seems to be happening is this: The hippocampus coordinates sensory and associative (interconnected) information received from many different parts of the cerebral cortex, and it stores that information. In this way broad, multisense memories (people's faces and voices; the look and aroma of a banana; the sights, sounds, and excitement of a circus) are stored in the hippocampus. Then, during sleep, portions of these hippocampal memories (like pieces of a jigsaw puzzle) are filed in various parts of the cerebral cortex and elsewhere, and ties between the hippocampal memories and these cortex memory pieces are created or strengthened. In this way, information gained during the last waking period is consolidated with previously stored cortex information of the same sort, making it easier for you to come up with new ideas and to work with this consolidated pool of information in the morning. Indeed, it seems likely that a prime purpose of sleep is to consolidate and file new information within the brain's vast data storage system, and that dreams are an incidental by-product of this process.

But of course, there is only limited room within the hippocampus. So gradually, over time, a given hippocampal memory is moved progressively, in disassembled form, to diverse parts of the cerebral cortex—leaving no information about this memory in the hippocampus or only enough to reconstruct the memory or an approximation of it should the need arise. This helps to explain why our conscious memories of past events tend to fade, even though we may be able to recall certain features of those events quite well.

We know this has a bearing on fear, anxiety, and the amygdala because the hippocampus has been implicated in fear conditioning. As Joseph LeDoux explains in *The Emotional Brain*, rats conditioned to a tone because they get a shock after it will also become conditioned to the background context, in this case their box. One can show this by removing the conditioned rats from the original box to another different box and then returning them some time later to the original box. The returned rats will show a fearful response to the old box, much as someone mugged in a particular alley will tend to feel anxiety if he or she later returns to that alley.

However, this context-specific conditioning is quite different from the rats' conditioned response to a simple tone. LeDoux and his colleagues showed this by conditioning rats to both a tone and their box. They then damaged the hippocampuses of certain rats. Those with no hippocampal damage continued being fearful of both the tone and the box, while those with hippocampal damage feared only the tone. This is presumably because the rats' response to the tone depended on sensory information received first from the thalamus and then (in a more refined form) from the cortex; but their response to the box depended upon complex contextual inputs received from the hippocampus. And so, according to LeDoux, "The amygdala is like the hub of a wheel. It receives low-level inputs from sensory-specific regions of the thalamus, higher level information from sensory-specific cortex, and still higher level (sensory independent) information about the general situation from the hippocampal formation. Through such connections, the amygdala is able to process the emotional significance of individual stimuli as well as complex situations. The amygdala is, in essence, involved in the appraisal of emotional meaning. It is where trigger stimuli do their triggering."[1]

The fact that the amygdala generates responses designed to save its owner explains why the amygdala is so much more hard-wired than most other brain regions. For you don't want flexibility here. You don't want to forget. You don't want to act neutrally or wonder what is happening when your life is threatened. You want to act. So the amygdala is set up to respond to progressively more sophisticated danger signals sent in by the thalamus, the sensory cortex, and the hippocampus. But it is not set up to be deprogrammed. It is not set up to forget. Nature has recognized the worth of remembering life-threatening experiences. So once some stressful situation has engraved a conditioned memory trace in the amygdala, that memory trace, in all probability, is there to stay.

This information provides a good theoretical groundwork for explaining how phobias develop and why they are so hard to undo. The amygdala has to start somewhere, so it probably starts with inherited neural patterns—patterns genetically primed to set up conditioned responses after very slight contact with things like suffocation, abandonment, falls from heights, poisonous insects, predators, and so forth. We know that the speedy information sent in directly by the thalamus is far broader, less specific, and less refined than the momentarily later information provided by the sensory cortex and hippocampus. So if, for whatever reason, this first pathway dominates the scene and sends in the strongest danger messages about, say, bugs and spiders, then it is easy to see how the amygdala could create hard-wired circuits generating a strong conditioned response to bugs and spiders, and also how these hard-wired phobic circuits could remain strong, broad, and entrenched long after all recallable hippocampal memories of any initial exposure to bugs or spiders have disappeared.

This anatomic setup can also account for a related anxiety problem called post-traumatic stress disorder or PTSD in which battle veterans or others who have experienced extreme trauma suffer from fear conditioning that causes them to periodically relive the terrors of that experience. For example, a Vietnam veteran exposed to shellfire might get a conditioned fear response that is set off by things like slamming doors—which would cause him to sweat, tremble, reexperience his battle terrors, and actually feel or act as though the old traumatic events were recurring.

Not all PTSD cases are so dramatic. Some merely involve recurrent distressing dreams or recollection of traumatic events, or else intense mental anguish or bodily arousal upon exposure to certain trauma-related cues. And PTSD does not require battle. Rather, it is more broadly defined as arising, at least in theory, from any severe traumatic experiences—such as personal assaults (fights, rapes, muggings, domestic violence, and so forth), as well as from disasters and accidents. This broad definition may well be justified. But PTSD has also risen to become one of the most fashionable of our current mental ills; and the net has been spread so wide that up to 30 percent of all war zone veterans are deemed to have it. According to *New York Times* writer Jane Brody, "an estimated 13 million people are disabled by this disorder"[2]; and even the relatively conservative NIMH estimates that "in the United States, about. . . 5.2 million people have PTSD during the course of a given year."[3]

We are quite suspicious of all this. We recall past occasions when the incidence of fashionable mental ills was vastly exaggerated. And we tend to agree with historian Edward Shorter, who thinks psychiatry may be shooting itself in the foot by overdiagnosing this disorder.* Even so, it seems clear that the number of people afflicted with PTSD or something like it is fairly large.

What makes PTSD different from phobias is the severely traumatic experience that brings it on. Here the brain is not prepared in advance, but the stimulus is so strong that it induces fear conditioning anyway—not only to the traumatic experience but also to a wide range of sights, sounds, and other conditions that go with it. These latter conditioned stimuli (like the sound of a distant howitzer) need not be strong; they are recorded in the amygdala because the unconditioned stimulus (the stress of battle) is so strong. And if these conditioned stimuli reach the amygdala mainly in the form of crude sensory messages relayed through the thalamus, then things quite different from booming howitzers (slamming doors, distant thunder) can be misinterpreted and can set off a conditioned fear response.

---

*One thing that makes PTSD hard to evaluate is the disability compensation associated with it. Specifically, potential benefits like removal from a war zone, hospitalization, lifelong financial benefits, and court awards can serve to promote simulation, exaggeration, and imaginary perpetuation of PTSD symptoms.

This certainly is not the end of the story. Among other things, researchers have uncovered abnormalities in what they refer to as the HPA (hypothalamic–pituitary–adrenal) axis. As shall be seen in the next chapter, the amygdala and hippocampus can send signals to the hypothalamus that end up causing the adrenal glands to secrete more or less of the stress hormone cortisol; and besides having various bodily effects, cortisol in turn influences the amygdala, hippocampus, and other brain regions in various ways.

Unlike victims of depression (who tend to have unusually high cortisol levels), people with PTSD tend to exhibit unusually low cortisol levels or show other evidence that the HPA axis is working oddly. Just how all this relates to PTSD is uncertain; but it seems likely that these abnormalities are important, and that their exploration will improve our understanding of what is happening.

Another common mental ill tied to this same arrangement is panic disorder. Panic attacks resemble attacks of phobia and PTSD, all three involving strong emotional arousal and activation of the sympathetic nervous system—producing hyperventilation, rapid heartbeat, sweating, trembling, and so forth. But panic has more subtle causes. Indeed, at least in the early stages, it gives the appearance of having no external cause at all. Instead, like REM sleep (which also activates the sympathetic system), panic attacks seemingly arise from internal signals, though they often give rise in time to various other sorts of conditioning and associated agoraphobia. All this makes these attacks harder to account for than simple phobic responses or PTSD attacks, and much harder for victims to avoid.

To get an informal feel for all this, one can hardly do better than consider a personal account by a lady we shall call Joyce Harper that appeared some years ago in the pages of the magazine *Good Housekeeping*.* Being a first-person report, the story contains no professional assessment and fails to do certain things like identify prescribed medications. Even so, it provides a good victim's-eye-view of the problem that illustrates how panic and associated agoraphobia do their work.

---

*Anonymous, "I Was a Prisoner of Panic," *Good Housekeeping,* June 1994.

Joyce Harper probably inherited a tendency toward panic disorder from her mother, who had a similar illness. At age 12 (in the mid–1970s) she herself began having panic attacks. She didn't know what they were, she just knew they weren't normal; so she kept them hidden as much as possible for fear of being labeled "crazy." This was especially hard in school, where inchoate terror gripped her periodically, preventing her from concentrating in class and causing her to respond to tests with a terrified blank mind, sweaty hands, and a racing heart.

She managed to graduate, and at her parents' insistence she learned to drive; but she found driving to be panic-provoking and traumatic. When an attack came in the car she would turn the radio up as loud as possible, because the noise somehow made it easier to cope.

After graduation she took secretarial courses at a nearby vocational college and got a part-time job as a clerk-typist. She could handle the job all right, but found the drive to and from work harrowing. When her boss asked her to run personal errands for other employees in a company vehicle, she refused. She then considered telling her boss the true reason for her refusal, but decided he wouldn't understand; so instead she gave notice.

In 1984, after being engaged for about a year, Ms. Harper married a man who knew about her problem. Thereafter, her life became progressively more restricted. She didn't get another job, she didn't socialize, and she quit driving. In 1986, after she and her husband moved back to her hometown, the panic escalated to a point where she never left the house. Even there she was not safe from waves of inexplicable terror—prompted by domestic events as minor as the ring of the telephone or the sight of a spider.

As this suggests, not all cases of anxiety end well. We have answers, as will be seen; but sometimes the victim won't seek help, isn't helped by therapy, won't stay on medication, reacts adversely to medication, or fails to respond to the drugs prescribed. For whatever reasons, it is fairly common to find people who are hideously crippled by panic disorder and associated anxiety problems like agoraphobia—being essentially confined to their own homes but feeling anything but safe because they are persistently overwhelmed by nameless terrors.

Also, most people with panic and other anxiety disorders are rational. No matter how terrorized, they seem able to cope with reality. So what they get by way of treatment depends very much upon themselves. This means that a good many victims who could be effectively treated stay away from treatment—because they are unaware of what can be done or else are afraid of being tarred with the stigma associated with mental illness. In addition, the treatment provided may be less than ideal because both the nature and the continuity of such treatment depend upon the felt needs of the patient.

Fortunately for the house-bound Joyce Harper, her husband heard about a panic disorder research program at a local hospital and enrolled her there. She got considerable relief from talking with professionals who understood her problem and meeting with people who had similar problems. She also received classic fear conditioning therapy that started by rating her fears on a scale of 1 to 10 and then had her confront the least fearsome five times a day—stopping when she felt anxious but always returning to try again.

This therapy seemed quite effective. In time she could walk to her mailbox and even (with her husband's help) go grocery shopping. But she felt she had a problem with the psychiatric drug that was part of the program. In her view this drug was causing her to lose weight, compromising her immune system, and making her continually ill. At this point she wanted a baby, and she worried about how the drug was affecting her body. So she quit the program; but she continued practicing the behavior technique she had learned, the technique of taking on lesser fears repeatedly until they became manageable.

Two years later, after giving birth to a daughter, she consulted a therapist for postpartum depression. The therapist, suggesting she had a chemical imbalance requiring medication, sent her to a psychiatrist; and the psychiatrist prescribed medication that besides relieving the depression also improved her ability to deal with panic. She found that she could again get to the mailbox, go shopping, go to church; and for the first time since her marriage she began making friends.

Things improved further when another psychiatrist prescribed a drug known to help specifically against panic. When Joyce Harper took this as indicated, she found she could enjoy all sorts of things she had never before tolerated as an adult—like eating in a restaurant, watching a movie in a theater, taking a leisurely shopping trip with friends, and going on increasingly long walks with her dog that took her up to a mile away from her home without panic. She still experienced considerable trouble driving—because worry that she might lose control during an attack tended to bring on panic—but in most other respects her lifelong disorder had been quelled.

While we aren't sure just what causes panic attacks, they appear clearly related to conditioning—only in this case the conditioning stimuli are not obvious external ones like snakes or the sounds of battle, but instead are internal stimuli (like rapid heartbeat and hyperventilation) or relatively gentle external stimuli—especially ones associated with prior attacks.

What allows these mild conditioning stimuli to provoke panic? Genes seem involved, because many panic cases like Joyce Harper's are inherited, and available evidence clearly indicates that genes play a major role. So in many or even most cases, the victim is genetically prone to develop the disease.

That makes it easier to see why internal signs of distress can promote panic. We know this can happen, because panic attacks can be induced by mimicking hyperventilation—a sign of bodily arousal that raises the level of carbon dioxide in the lungs. Specifically, investigators have found that having subjects inhale a gas mixture rich in carbon dioxide can sometimes provoke panic attacks—and that such attacks are more likely to occur in people with panic disorder than in normal subjects.

This seems logical, because if phobics and PTSD victims can become conditioned to a broad swath of external stimuli ranging from cockroaches to banging doors, then there seems no reason why panic sufferers cannot be conditioned by hyperventilation or other internal stimuli associated with sympathetic nervous system arousal.

Even so, we have not yet penetrated to the heart of this mystery. We know that the amygdala at the core of the conditioning process has reciprocal ties to the brain stem centers that govern breathing. But it also has a rich array of connections to other areas including the prefrontal cortex, the hippocampus, and the locus coeruleus—this last being a brain stem region whose stimulation can provoke panic.

So there appear to be various possibilities. Among them: It appears conceivable that hyperventilation (perhaps associated with feelings of suffocation) and other internal stimuli could be responsible for the initial "panic" conditioning. But it seems at least as likely that people prone to the disorder are being conditioned by early stressful situations like being away from home (in school) or taking tests, much as people predisposed to a snake phobia can have this phobia activated with little or no actual exposure to poisonous serpents. In that case, hyperventilation and other internal stimuli would be secondary rather than primary causes whose association with past attacks could generate sufficient worry to prompt new attacks.

In any event, things don't end here in most cases, because the brain of a panic victim does not limit its associations to internal physical stimuli. Instead, it commonly associates panic with a widening array of external triggers like driving a car or encountering strangers. That is, once some new event has evoked worry or been associated with an attack, the victim may develop a conditioned response to that event. In this way the list of triggering events grows, sometimes leading to agoraphobia (fear of situations from which escape may be difficult or embarrassing) and making the victim increasingly house-bound and reclusive. Of course, as Joyce Harper discovered, seclusion does not solve the problem. Instead, triggering events arise inside the home; and so the victim becomes increasingly crippled by inchoate terrors arising from such mild domestic stimuli as mice scampering in the walls or a pot boiling over on the stove.

So far we have been talking about anxieties that seem directly tied to a conditioned fear response. But there are other anxiety disorders, the two most common being obsessive-compulsive disorder (OCD), in which people irrationally repeat certain acts, and generalized anxiety disorder (GAD), which is not stimulus-provoked but which involves levels of anxiety ranging from marginally tolerable to absolutely incapacitating.

OCD can involve repetitive actions like washing, cleaning, checking, praying, or putting things in order; but it can also involve obsessive thoughts commonly tied to unreasonable doubts (Do I have AIDS? Have I harmed someone on the road? Did I lock the door?). Though OCD runs the gamut from being barely perceptible to extremely debilitating, those afflicted typically spend hours each day dealing with their compulsive thoughts or habits. They know these behaviors are "crazy" and in social or other situations can suppress them for a time; but they cannot control them indefinitely, and they also feel these actions are needed to stave off danger or discomfort.

Interestingly, we have evidence that the obsessive repetitions of OCD have something in common with repetitive magical and religious rituals in various cultures. OCD also shares a genetic heritage with Gilles de la Tourette's syndrome (a syndrome of mostly facial and vocal tics progressing to outbursts of repetitive speech and foul language), because people with one disorder often have the other; the families of OCD patients have higher than normal rates of tics or Tourette's syndrome; and the families of Touretters have higher than normal rates of OCD.

Research evidence suggests that various deep cerebral structures including the amygdala, hippocampus, and caudate nucleus (see Figure 5.2), are involved in OCD. We know that these structures receive information from the cortex, but then project it back via the thalamus, primarily to the cortex's frontal lobes. Brain imaging studies indicate that people with OCD have abnormal neural circuits running between a small part of the frontal lobes (the orbito-frontal cortex) and portions of deep cerebral structures—especially the caudate nucleus. So it is possible that the repetitive thoughts and acts of OCD arise from failure to shut down activated neural circuits in this path.

The last anxiety disorder on our list, generalized anxiety disorder (GAD), is characterized by excessive worry and anxiety afflicting the victim on most days for a period of some six months or more. Besides being hard to control, the worry tends to go hand in hand with things like restlessness, fatigue, irritability, muscle tension, poor concentration, and disturbed sleep. Though the illness often occurs in combination with other anxiety disorders, the victim's worry is not exclusively directed at any defining features of those ailments, such as worry about having another panic attack in the case of panic disorder.

Many people (an estimated 4 to 6 percent of the entire U.S. population) are said to develop generalized anxiety disorder sometime in their lives, so the illness clearly deserves attention. But its classification is odd, being based on the excess worry and anxiety remaining after all other causes have been accounted for; and it is so commonly associated with other anxiety disorders or depression that some experts have questioned whether it really exists as an independent entity at all.

If we suppose that it does (the issue has not been decisively resolved), that leaves us pretty much in the dark as to this disorder's cause. But since the ailment seems rather vaguely defined at this point, a somewhat vague explanation is the best we can expect.

In general, current information suggests that GAD, like many other anxiety disorders, involves strong emotional stimulation of the frontal cortex by the deep cerebral structures—a pattern consistent with the fact that many people afflicted with this disorder also have depression or other anxiety disorders like phobias or OCD. We aren't sure how this stimulation is accomplished; but it seems possible that it is brought about partly through generation of acetylcholine by the nucleus basalis, a gray matter nucleus controlled by the amygdala. In addition, it seems likely to involve serotonin, norepinephrine, and other neuromodulators coming up from the brain stem. In general, it seems likely that most anxiety attacks generate a high-pitched waking state, and that the insomnia often associated with these disorders arises from levels of neuromodulators (notably serotonin and norepinephrine) and levels of brain activity that are too high to be compatible with sleep.

So where do we stand with regard to treatment? Anxiety disorders are naturally persistent, mainly because the amygdala is so hard-wired. Even so, as Joyce Harper's case shows, we do not lack for answers.

To begin with, we have a large body of moderately to highly successful psychotherapy. For more than a century, psychologists and psychiatrists have sought treatments for anxiety disorder patients—of which those with phobia, PTSD, and panic have constituted the lion's share. The ailments subjected to such therapy in times past had different names, like "hysteria" for example, but by and large they were the same anxiety disorders we know today. And since past efforts to treat other leading disorders (notably depression, bipolar disorder, and schizophrenia) were typically less successful, anxiety has long been the mainstay of psychotherapy.

So what has psychotherapy devised? As one might imagine, there are all sorts of approaches, the main ones involving psychoanalysis, behavior therapy, or cognitive therapy. By and large the behavioral ones seem to have been the most productive. For even though behaviorism looks a bit old-fashioned now, we really are dealing with specific hard-wired conditioned responses that cause most anxiety disorders. And since the behaviorists made conditioned responses their bread and butter, it's not surprising that behavioral therapists should appear to have made the largest gains.

The classic behavioral "desensitizing" method was devised by Joseph Wolpe. We know that rats conditioned to fear a tone (because the tone was followed by a shock) can be "desensitized." That is, if the rats are moved to another box, and if the tone is then sounded from time to time without administering any shock, in due course the rats will lose their fear of it. Seeking to apply something like this to people, Wolpe would get a patient with a phobic disorder into a relaxed state, work up a list of stimuli ranked according to their ability to provoke arousal in that patient, and have the patient envision the weakest stimulus on the list while in the relaxed state until he or she could do so comfortably. Then they would go on to the next item. Meanwhile, between appointments, Wolpe would have the patient seek out harmless and manageable anxiety-provoking stimuli in order to show that they were harmless—thereby reinforcing the whole process. This approach has been quite successful in extinguishing conditioned signals that prompt anxiety attacks.

Behavior therapy also appears to have made modest gains against OCD. In this case the patient is exposed to some stimulus (dirt, germs, etc.) that provokes the compulsive behavior, but is told to refrain from it. This treatment is emotionally trying, and success is generally incomplete, so it is best to combine it with drug therapy, which usually involves the tricyclic antidepressant clomipramine or one of the SSRIs.

Assuming this limited success against OCD is genuine, we have no precise explanation for it. However, brain mechanisms are state-dependent;

psychotherapies (including behavior therapy) can influence brain states; and so we can see how behavior therapy might help against OCD.

Another variant on this technique is used to deal with the so-called "social" phobias that cause those afflicted to fear crowds or people. Anxiety about approaching other people is easy to understand, because it's hard to tell what other people are going to think or how they will react; and in fact it's quite common for anxiety about this to reach phobic proportions. But if you have a protected group setting where everyone agrees not to criticize or "tell" on anyone, you can start playing games like having everyone write anonymously on a slip of paper the worst thing that anyone could think of them; and when the slips are pulled out of a hat and read they usually all say about the same thing . . . "they'll think I'm a homosexual," or "they'll think I'm a terrible person." This helps to show that these fears are universal, and so everyone starts to breathe easier and to talk to one another, and in time they can go out on assignments to talk with other people and come back and report on how that went, and so on. It's a kind of social behavior therapy, and it works.

But behavioral desensitization has an Achilles' heel. For it turns out that fear conditioning "extinguished" through desensitization can also be revived. Consider the desensitized rat cited above. If this animal is returned to its old box (where it came to fear the tone), it is likely not only to fear the old box but to fear the tone again, despite the fact that no shock follows. For the amygdala is hard-wired. We cannot change its circuits. The best we can do through therapy is to change circuits elsewhere, mainly in the cerebral cortex, that provide the amygdala with information. In the case of behavioral desensitization, we may be altering associations that cause the cortex to send particular information on to the amygdala. In other words, we may be erasing or modifying "danger" signals within the cortex. But the amygdala also receives information from other sources, including the hippocampus and thalamus; and so, if old associated stimuli get passed along by them, or if the amygdala gets aroused for other reasons, the old wiring in the amygdala may react, reestablishing pathways to the "extinguished" danger markers in the cortex and reviving the old conditioned fear response.

One can try to deal with this by making patients consciously aware of what is happening and getting them to grapple consciously with the problem. That is essentially what the Freudians and cognitive psychotherapists say they are doing—the Freudians through a cathartic exploration of old memories and the cognitive psychotherapists through a more direct appeal to the power of reason.

Neither approach should be dismissed; but getting the thinking parts of the brain to exert conscious control over the amygdala's fear machine isn't easy, partly because the cortex's connections capable of exerting some control over the amygdala are far weaker than the amygdala's connections to the cortex, the nucleus basalis, and centers in the brain stem and elsewhere that evoke an emotional response. Perhaps for this reason, it has been claimed that the Freudians are not really appealing to the patient's consciousness alone. According to this argument, by getting the patient relaxed on a couch and talking about anxiety-provoking matters they are really engaged in something akin to desensitizing behavior therapy.

They may also be enlisting the placebo effect. As we have seen, a rat conditioned to fear a shock following a tone will produce a panicky, stressed-out response at the tone, even if no shock follows. But if it knows it can avoid the imagined shock by hopping up onto a board, it will jump onto the board at the tone while exhibiting little or no stress. It knows it has control over the problem; and this knowledge, conveyed by other parts of its brain to the amygdala, dispels its conditioned fear.

Today we tend to have a lot of faith in pills. So if a doctor prescribes a sugar pill for a patient and says it will probably cure the patient's problem, the patient feels that he is in control. The pill is very much like the rat's board. It permits the patient's brain to reassure the amygdala, defuse stress, and jack up the body's immune response to a wide range of ailments. More specifically, pretty nearly everyone agrees that psychiatric drugs' effectiveness against stress-related problems (fear-conditioned anxiety disorders and depression) is due partly to the placebo effect. And while we see no justification for occasional claims that the placebo effect does everything and our drugs do nothing, it seems clear that the placebo effect is important.

But placebos do not have to take the form of pills. They can also take the form of any treatment convincing to the patient. So aside from their other benefits, it seems clear that a wide array of treatments—drug prescription, psychoanalysis, behavioral therapy, cognitive therapy, and so on—can engage the placebo effect and in this way provide genuine benefits. In the case of most anxiety disorders these benefits will not be decisive because the amygdala is hard-wired; but they should certainly not be ignored; and since the placebo effect is strengthened by continuity of care, this is one more reason to emphasize the need for continuity.

It could also be one reason why nearly all studies done to date show that anxiety disorders (like most mental ills) are best treated by a balanced blend of therapy and medication rather than by either one alone. The trick, of course, is knowing where to strike the balance—something that needs to be decided for each patient by someone familiar with the case, acquainted with its current course, and experienced in both drugs and therapy.

Within this context, something unusual about the anxiety disorders should be noted. That is, they appear to respond favorably to many antidepressants—not just to the early tricyclics like imipramine and the addictive benzodiazepines but also to the SSRIs. That's a bit surprising, because while serotonin levels tend to be subnormal in depression (so that boosting those levels with an SSRI could theoretically prove beneficial), the anxiety disorders do not typically involve low serotonin levels. If anything they involve *high* serotonin levels. So making those levels higher still would hardly be expected to calm things down, which leaves us with something of a mystery as to how these drugs achieve their desirable effects.

One thing the SSRIs probably do is activate the front of the cerebral cortex (notably the frontal lobes) at the expense of the amygdala and other deep cerebral structures. That's because they boost serotonin, serotonin boosts the front part of the cerebral cortex, and so the SSRIs may boost the cortex's limited ability to control what the amygdala is doing.

They may also be muffling the amygdala. It turns out that the amygdala taps a nearby gray matter nucleus, the nucleus basalis in the forebrain, and tells it to release the neuromodulator acetylcholine when the amygdala wants to arouse the cortex. We know this is an essential step in the amygdala's arousal task, because artificially stimulating either the amygdala or the nucleus basalis arouses the cortex; and more dramatically, blocking the action of acetylcholine keeps the cortex from being aroused by conditioned stimuli, or by an artificially stimulated amygdala, or by an artificially stimulated nucleus basalis. And since boosting serotonin tends to put the brakes on acetylcholine, the SSRIs may be restraining the amygdala and other deep cerebral structures as well by working against the acetylcholine system.

In a way, though, all this is just part of a larger puzzle. For the SSRIs seem to be making headway against a wide range of mental problems, including the sadness of depression and the discomforts of minor stress. They don't seem addictive like the benzodiazepines or heavily burdened with major side effects like the tricyclic antidepressants. They can make us feel more alert and sharper, as those who take them for cosmetic reasons have discovered. And one of their more questionable cohorts, the recreational drug ecstasy (methylenedioxymethamphetamine or MDMA), can induce prolonged euphoria by pulling a big lever, releasing virtually all the serotonin in the system at one time, and preventing neurons from reabsorbing the serotonin they release.

We don't know precisely what to think of all this. That is, we don't know precisely what to think of the SSRIs. In the case of the drug ecstasy, anybody should be able to see that drenching the brain with serotonin is unwise. We have little conclusive evidence of harm, partly because widespread use of ecstasy is new. There are tentative indications that ecstasy damages neuron receptors, causes depressive aftereffects, and impairs memory. But we really don't need these tentative indications. We know that serotonin is a powerful neuromodulator, vital to proper brain function, and that the brain is not designed to cope with the serotonin flood that ecstasy releases. So a strong downside here is very likely.

The SSRI situation, as we have seen, is very different. Here the brain is not being grossly abused by binge-type exposure to a party drug. Instead, its long-term modulatory balance is being changed. We have strong evidence that this can help deal with a broad range of mental ills, and we are not sure why that is. It may be, as noted above, that raising serotonin levels tends to favor the frontal (thought and judgment) part of the brain over the emotion centers deep in the cerebrum. And since the lion's share of mental ills involve emotional problems, the SSRIs may help cope with such problems while at the same time enhancing the waking state and making everyone feel better.

But there is an evolutionary mystery here. If serotonin can make us function better by relieving a host of mental ills and improving our performance, that would seem to promote survival. So why hasn't Nature seen this? Why hasn't she set our serotonin level higher? After all, she has had millions of years of human and prehuman brain evolution in which to do so. Is it simply that we live at a higher level of stress today and so need more serotonin than we did in ages past? Or is it that we are missing something, that there is more of a downside than we see, something that might not be self-evident or that could take a long time to appear?

We don't know. Nobody does. Perhaps whatever the ecstasy users run into will help us get a handle on all this. Meanwhile, the SSRIs and certain other drugs have a proven track record against many severe anxiety disorders. So saying "no drugs" is wrong, just as saying "no therapy" is wrong. Rather, we should seek a balanced approach; and since therapy looks both safe and effective, it seems like the best way to start off, especially when the disorder seems relatively mild. That leaves the drugs, with their relatively powerful effects, side effects, and unknown long-term consequences, for use when therapy alone has failed or seems unlikely to yield a satisfactory result.

9

Searching for Doctor Doom

To sit in solemn silence in a dull dark dock,
In a pestilential prison with a life-long lock. . .

—W. S. Gilbert, The Mikado, 1885

What is depression? It's not just sadness, because all sorts of things from lost soccer games to failed romances to death of a loved one can make us feel sad in ways that are perfectly healthy and normal. But if the gloom has no obvious cause or goes on and on for a clearly abnormal length of time, it crosses into the realm of mental illness. This class of mental illness, called depression, takes an enormous toll in terms of personal happiness, economic productivity, and human life.

It turns out that the depressive disorders tied to irrational sadness have a lot in common with the anxiety disorders tied to irrational fear. Both are mysterious emotional problems. Both involve stress. Both work through many of the same brain structures, processes, and pathways. Both can be helped by a combination of psychotherapy and drugs. Both have many different variations. And both defy easy scientific analysis or statistical assessment.

The statistical assessment problem arises mostly because both sadness and anxiety are useful feelings. Nature tells us that we should worry about some things and feel sad about others. So at the lower end of the disease spectrum it is hard to tell what are exaggerated but still normal responses and what are mild cases of mental illness. This is not the kind of situation that pleases statisticians.

But pleased or not, statisticians keep churning out numbers. That's their job. So, according to the 1999 mental health report of the U.S. Surgeon General, the best available data show an annual prevalence of depression among all U.S. adults of about 7 percent. Though this rough figure was well below the report's 16 percent prevalence estimate for anxiety disorders, it was still impressive—especially since depression tends to last, creates substantial disability, can involve psychosis, and can prove life-threatening by raising the risks of heart disease and suicide.

The three kinds of depression included in this 7 percent are major depression, dysthymia (a chronic form of depression), and bipolar disorder (depression alternating with mania). While the last two certainly merit our attention, major depression without mania accounts for something like three-quarters of these cases. Victims typically show reduced interest in most activities, get reduced pleasure from them, and have a depressed mood most of the time. Other common symptoms include chronic insomnia (or, conversely, chronic oversleeping); unusual weight loss (or unusual weight gain); fatigue; lessened ability to think, focus, or make decisions; feelings of worthlessness or guilt; and recurrent thoughts of death, with suicide being a serious danger. A single episode of untreated major depression will typically last something like 9 months. According to a recent study by the World Health Organization, the World Bank, and Harvard University, major depression is the leading single cause of disability in both the U.S. and the world, producing direct and indirect costs in the U.S. alone of over $30 billion per annum.

Dysthymia tends to be less severe than major depression but lasts indefinitely. People with dysthymia can have episodes of major depression, but unlike these bouts of major depression the dysthymia doesn't often go away without treatment.

Bipolar disorder afflicts between 1 and 2 percent of all Americans. The most bizarre of the three kinds of depression, it puts the victim on a roller coaster, alternating depressive episodes with periods of normalcy and periods of mania—the average victim having a manic episode once every two to four years.

Mania, which can involve elation, euphoria, grandiosity, paranoia, psychosis, and grossly impaired judgment, has legendary home-wrecking and job-wrecking power. There are forms of bipolar disorder in which the manic side is mild and does not match the clinical definition of mania. But that's the exception rather than the rule. Most cases come with a high energy and high libido that tend to promote wildness—spending sprees, recklessness, promiscuity, and other kinds of offensive or uninhibited behavior. Jonathan Leonard recalls the all-too-common example set by a realtor his parents knew who one day left home and took off across the country on a string of credit cards—thereby losing his savings, family, and business to this illness. Not surprisingly, the morning-after view of what happened in the manic state isn't pretty, which helps to explain why up to 20 percent of all manic depressives commit suicide.

It is also noteworthy that some manic depressives attract attention not just because they behave oddly but because they are well known. It turns out that an optimistic, upbeat, slightly manic approach to life is a help in certain professions like real estate, marketing, and acting. So it is not surprising to find that a fair share of people in such professions are slightly manic; or that some have severe mood swings; or that some of these exceed the bounds of rationality, entering the grim bipolar world of manic psychosis and suicidal depression.

For instance, consider the case of Lois Lane—not the fictional Superman heartthrob but the flesh and blood actress Margot Kidder, who during the 1970s and 1980s played that role in four Superman movies.*

Kidder was born in 1948 at Yellowknife, a raw mining town in Canada's Northwest Territories. Her father was a mining engineer whose job took him to such places, so Margot spent her childhood in this and

---

*Margot Kidder's case was widely reported in the press. See J. D. Reed, *People Weekly*, September 23, 1996; B. D. Johnson, *Maclean's*, May 6, 1996; P. Lambert, *People Weekly*, May 6, 1996; and D. Thigpen, *Time*, May 6, 1996.

similar locales. Eventually, as she blossomed into a gorgeous teenager, she was sent away to boarding school—partly to protect her from the rough ways of the mining community—and it was there she became fascinated with acting.

It was also there that she began having severe mood swings. (At age 14 a breakup with a boyfriend prompted her to swallow a handful of codeine pills.) She then proceeded to finish up at her boarding school two years early, moved to Los Angeles when she was 18, and began landing acting jobs.

Three years later, in the 1960s, she began seeing psychiatrists for her mood swings. These were peak years for psychoanalysis. However, she came to distrust the prevailing psychoanalytic methods—which encouraged her to express emotions—because she thought such encouragement was absurd for someone like herself whose problem was a surfeit of emotions.

Margot Kidder acted in a string of movies, including *The Great Waldo Pepper,* before landing the Lois Lane role in the first of her Superman movies. That was in 1978. Ten years later a psychiatrist diagnosed her as having bipolar disorder, but her major ups and downs had started long before and in fact had become notorious. In the course of her career she went through three marriages and a long string of boyfriends—including former Canadian Prime Minister Pierre Trudeau, *Superman III* costar Richard Pryor, and *Jurassic Park* author Michael Crichton. Throughout, she was only too aware of her underlying emotional instability, referring at one point to her life as "grand and wonderful, punctuated by these odd blips and burps of madness."[1]

In 1990, while working on a set for the *Nancy Drew and Daughter* television series, she was in an auto accident that injured her back and confined her to a wheelchair. A later operation restored her ability to walk, but $600,000 in medical bills drove her into bankruptcy, and pain encouraged addictive attachments to pills and alcohol. "In those days," she said, "if I felt myself starting to go manic, I'd get drunk. Better drunk than crazy."[2]

In 1996, after taming her substance abuse, Kidder was working on her memoirs in a creatively manic state, writing up to 12 hours a day, when a computer glitch gobbled up her text. Seeking to retrieve three years of work, she flew with her laptop computer to Los Angeles on April 16th, only to be told by a data retrieval company that her files were truly lost.

She had been scheduled to fly on to Eastern Arizona College and give a class in career management for actors, so she returned to the Los Angeles airport four days later. By then, however, manic psychosis and the paranoia commonly associated with it had set in. She became convinced that her first husband (Thomas McGuane) and the CIA wanted to kill her to prevent her book from being published. And she began seeing conspirators and assassins everywhere.

At 3 A.M. she was still in the airport, shouting at people and telling a traveling television crew that her husband wanted to kill her. At that point she looked tired and dirty. She tried to take a cab, but found she had no money. She then tried to use her ATM card, but ran off in fear that the ATM machine was about to explode. She also threw away her purse, believing it contained a bomb.

On April 21st, having slept where she could, she made her way to downtown Los Angeles, some 20 miles from the airport, where she was taken in by a street person named Charlie who was living in a cardboard box. He took reasonably good care of her; but another man tried, apparently unsuccessfully, to rape her, dislodging the last of several caps on her front teeth in the process.

The next day she tried to reach a friend's house some 12 miles to the north. By then she had cropped off most of her hair to disguise her appearance. That night she slept in a motel room provided by some Alcoholics Anonymous members. The next day her paranoia appeared to have abated, for she wandered into the backyard of a local residence and told the startled homeowner that despite her appearance she was Margot Kidder.

The homeowner called the police, who found Kidder and took her to the nearby Olive View Medical Center for observation. Fortunately, Kidder had five siblings and a daughter interested in her welfare who were willing to provide a support net. Her sister Annie, a Toronto theater director, became aware of Margot's plight and had her transferred to mental health facilities at the UCLA Medical Center. The doctors who saw her at UCLA appeared first to have been concerned about her safety, for she was not im-

mediately discharged upon request. Instead, her release was delayed until
April 30th, when a judge ruled that she was no immediate danger to herself
or others.

To avoid the press, Ms. Kidder stayed at a rented house that her family obtained for her on an island near Vancouver. There her brother John, a Vancouver inventor, introduced her to literature on bipolar disorder. This apparently jolted her into recognizing her problem and accepting its
diagnosis. So she proceeded to take Depakote, an antiseizure drug then recently approved by the Food and Drug Administration for short-term control of manic depression, and her disorder proceeded to settle down.

Since then, Kidder has picked up the pieces of her life. She found a computer expert who was able to recover the lost text files for her book. And despite Hollywood's edginess about any mental state that might prevent com-
pletion of a film, she has managed to secure strong secondary roles in two
films (*Hi-Line* and *The Annihilation of Fish*) that were released in 1999.

This case has a number of unusual features, obviously including the victim's prominence and strong character. But it also exhibits various features
that are fairly common. Among other things, the strength of her periodic
mood swings varied greatly. Indeed, the power of those swings gives every
appearance of depending on an array of internal and external circumstances; some of the swings appear to have been minor; and so far as one
can ascertain from published records, it was only the last swing that produced truly catastrophic psychosis.

Also, like many manic depressives, Kidder appears to have enjoyed her
bouts of subpsychotic mania. They allowed her to work long and intense
hours on her book. They made things like attending symphonies and many
other facets of her life more vivid. And they produced a heightened sense of
euphoria, accomplishment, and enjoyment that was simply not available at
other times. Like many other bipolar victims, this made her reluctant to accept treatment—or even to acknowledge the correct diagnosis of her ailment—until she had lost all control, pitched over the crest of euphoric mania into frank psychosis, and actually struck bottom.

Like many mental ills, mania and the various forms of depression have both a genetic and an environmental side. Because of the genetic component, close (first-degree) relatives of people with these mood disorders run two to four times the normal risk of contracting a mood disorder themselves, the risk being especially great if the relative has bipolar disorder.

On the environmental side, we know that various kinds of stress—including parental neglect, physical or sexual abuse, and other forms of maltreatment—promote depression. This is especially true if the stressful situation is prolonged. Also, certain medical ailments (most notably stroke) can cause depression. And certain temperaments can play a role, because someone with an easygoing personality is markedly less prone to depression than someone who is high-strung and neurotic.

Regarding treatment, we know that various treatments for depression and associated mania have improved to a point where the biggest single problem today is getting afflicted people to come in for treatment and stick with it. Naturally, since depression reduces motivation, someone who is depressed tends to have scant motive for starting or pursuing treatment, while someone who is manic like Margot Kidder is erratic and tends to have all kinds of better answers. Also, primary care physicians sometimes fail to recognize depression; the social stigma associated with mental illness may keep depressed people from coming in; money and health coverage can be issues; and not everyone realizes that treatment can yield strong benefits. So it is not surprising to find that most people with depression receive no specific form of treatment, or that many with bipolar disorder go untreated for long periods.

That's regrettable, because treatment is good and getting better. All by itself cognitive-behavioral therapy, which came of age in the 1970s, can produce strong benefits. This therapy recognizes that depressed people commonly have a negative view of themselves, the outside world, and the future. So it starts out by setting up a warm and supportive therapist–patient relationship. Within this relationship, the therapist applies logic, rules of evidence, and Socratic questioning in ways that encourage the patient to reveal, question, and correct the assumptions responsible for his or her bleak outlook. Interpersonal therapy, which takes a less structured approach, also features face-to-face discussions with the therapist and also has a good track record. These therapies, generally involving six to twenty weekly visits, are preferred by many over drug therapy for dealing with mild to moderate cases of depression.

Meanwhile, the drugs have become diversified. A wide range of drugs are available—including the older tricyclic antidepressants (TCAs), monoamine oxidase inhibitors (MAOIs), and selective serotonin reuptake inhibitors (SSRIs). All can have significant problems and side effects. Even so, most of these drugs seem to produce initial positive response rates at least as good as those obtained with psychotherapy and better results in cases of severe depression. The initial course of treatment is somewhat shorter (typically six to eight weeks of weekly or biweekly visits to provide support, monitor side effects, assess symptoms, and adjust doses). Antidepressant therapy should be maintained for at least six months to one year, and recurrent depression may require lifetime therapy. Unlike psychotherapy, drugs can sometimes subdue severe cases; when used to supplement a mood stabilizer (usually lithium, carbamazepine, or divalproex) they may work against bipolar disorder; and when used together with antipsychotic drugs they can help bring depressed patients with psychosis back to rationality.

For possibly suicidal, clearly suicidal, or psychotic patients, the drugs' biggest drawback is the two or three weeks they need to take effect. That's becauce the barbaric state of our current laws and HMO regulations can make it hard to keep patients who are intent on suicide or mayhem in a hospital or any other secure place long enough to ensure that a chosen drug is doing its assigned job. But there is an alternative, one that experience has shown to provide a reliable and quicker fix for depression with fewer significant consequences than any of the antidepressant drugs. That alternative is electroconvulsive therapy, otherwise known as shock therapy.

Shock therapy works by passing an electric current through the patient's brain, thus overactivating multitudes of neurons and causing them to convulsively discharge—as in an epileptic seizure. Among the affected neurons are ones in the brain stem that are responsible for distributing serotonin and norepinephrine to other parts of the brain. These particular neuro-modulators are thought to be depleted or less than normally effective in depression. Shock therapy reverses that. This may not be the only mechanism operating here, but it is an important mechanism, one that helps to explain why four to eight shock treatments given on successive or alternate days can temporarily dispel depression and elevate the patient's mood.

Shock therapy is not popular, partly because it has been successfully vilified in the 1975 film *One Flew over the Cuckoo's Nest* and elsewhere, and partly because nobody conditioned to fear electricity from childhood can calmly contemplate passing the equivalent of house current through a human brain. But contrary to what common sense tells us, long decades of experience have shown this procedure to be highly beneficial for a good many depressed patients as well as essentially harmless when used in moderation. Thus, if the informed consent now universally required can be obtained from the patient, and until better laws and health rules governing temporary hospitalization are established, shock therapy will remain the best way of securely restoring rationality to severely depressed people who are suicidal or psychotic.

I (Allan Hobson) was astonished when a depressed outpatient of mine calmly demanded shock therapy and firmly refused pills because he wanted quick and effective relief. He got it. After three shock treatment sessions at two-day intervals, he said "That's enough!" And it was.

Turning back to the vast majority of less dramatic cases, we find that not all patients respond to psychotherapy, nor do all respond to drugs. Thus, there seems good justification for using an integrated combination of the two. But that is only part of the answer, because regardless of whether we use drugs or psychotherapy or both, we are dealing with long-term problems that have a notorious tendency to relapse. Thus, after the initial treatments have substantially reduced the initial symptoms, the patient should receive at least six months of additional therapy (usually consisting of biweekly or monthly visits) to prevent relapse and ensure elimination of all symptoms before treatment is tapered off. Finally, maintenance drug therapy (including monthly or quarterly visits) should continue indefinitely for patients who have had three or more prior episodes of the disease, regardless of whether this is major depression, chronic depression, or bipolar disorder.

That is the bad news. The good news is that brain science is beginning to get a better handle on depression. The full story is still wreathed in mystery; but we have a lot of clues that implicate all sorts of things—the amygdala, stress, body chemicals, assorted neurotransmitters, the prefrontal lobes, sleep patterns, and the thalamus. Some of these clues lead from one thing to another. So right now this hard science research has a bit of the feel of a kids' treasure hunt, where notes found at one site tell where to proceed next, and also guide the searchers toward knowledge about how to improve treatment.

Those seeking a bird's-eye view of this quest should start with an awareness of the hypothesis that depression goes hand in hand with chronically high sensitivity to acetylcholine, and also with subordination of the thought and judgment centers in the frontal and prefrontal cortex to the amygdala and its emotion-driving companions deep in the cerebrum.

The first logical way station in this hunt is the amygdala. As we saw in the last chapter, the circuits in the amygdala that respond to danger have vast power. Besides projecting strongly to the cerebral cortex, hippocampus, and brain stem, they can also order the nearby nucleus basalis to further rouse the cortex by sending out a flood of acetylcholine. So here, in the fight-or-flight response, we have a picture closely resembling that found in sadness and depression.

That may help to explain why roughly half of all depressed people also have anxiety disorders—because the same kinds of brain imbalances cause both conditions. But it still seems very strange. We can easily see how anxiety disorders (inappropriate fear or worry) can result from inappropriate programming or triggering of the amygdala. But the amygdala's fight-or-flight response rouses the whole brain and body. So how does this very strong positive response get flipped to produce the negative responses of sadness and depression?

If we try to imagine the prehuman and tribal times when sadness was evolving, we can see that sadness likely had survival value. Even today, members of an animal group or isolated primitive tribe who are not the alpha male or tribal chief will find that feistiness can provoke violence, wounds, and sometimes death. So unless one is seriously contending for leadership or status, subordination pays. Thus, a degree of sadness following defeat, one that encourages acceptance and resignation, can promote survival.

We are hardly the first to realize this. Charles Darwin, in his book *Expression of Emotion in Man and the Animals*, emphasizes the value of appeasement gestures by defeated combatants. In general, he noted that where confrontation begets aggression, surrender promotes peace.

In a similar vein, death of a spouse, parent, or close friend often means a big change in social status. Most societies never reached the Indian extreme in which widows were encouraged to throw themselves on their husband's funeral pyres. But even so, we can see how bereaved members in a small group of apes or hunter-gatherers would be well advised to keep a low profile for a time, consistent with their uncertain social status. And obviously, such a low profile is strongly encouraged by the intense sadness we associate with grief.

But this only shows the obvious—that sadness and depression had reasons to evolve. It still doesn't explain how the rousing fight-or-flight response got flipped into sadness. To explain that, even in theory, we need to imagine a mechanism in the brain that could produce this melancholy but desired result.

The most likely mechanism is prolonged stimulation. Time after time, in looking at neurons and the brain, we find that stimulation causes arousal, but that excess stimulation causes dysfunction or even damage.

For instance, look at what the amygdala does to the hippocampus. When the amygdala senses danger, it issues chemical orders that kick off a cascade of responses. One of many ways it does this is by telling the hypothalamus to release a chemical called corticotrophin-releasing factor (CRF). This starts a chain of activity that causes the adrenal glands atop the kidneys to secrete stress hormones that help gear the body up for action. Some of these stress hormones travel back through the bloodstream to the brain, where they bind to receptors on the amygdala, the hippocampus, and elsewhere. The hippocampus then responds by telling the hypothalamus to stop releasing CRF. But so long as the amygdala senses danger it will keep telling the hypothalamus to release more; and the balance between encouragement on one side and discouragement on the other will ultimately determine how much CRF and stress hormones get released.

So long as the real or imagined emergency is short-lived, well and good. The amygdala shuts off, the hypothalamus obeys the hippocampus, and the flows of CRF and stress hormones stop. But if the stress goes on too long, the stress hormones overstimulate the hippocampus, the hippocampus starts to falter, its ability to oppose the amygdala lessens, and its capacity to perform its other duties is impaired. In theory, repeated episodes like this could do major long-term harm by shrinking the hippocampus, causing memory problems, producing chronically high levels of circulating stress hormones, actually killing neurons in the hippocampus, and causing the amygdala to become chronically hyperactive.

We have evidence that all of this actually happens. Researchers have found that socially stressed mice and monkeys undergo marked shrinkage of the hippocampus. Likewise, people suffering from a condition called Cushing's disease, which generates excess adrenal stress hormones, experience both memory problems and hippocampal shrinkage, as do certain victims of post-traumatic stress disorder.

But the really interesting thing is that this same process applies to depression. People with depression tend to have a smaller than average hippocampus, and the degree of atrophy tends to be greater among those who have been depressed longer. Conversely, the amygdala is more active than normal in both sadness and depression, and this overactivity persists even after the depressed state has passed. Finally, the brains of many depressed patients contain abnormally high levels of adrenal stress hormones. So it seems clear that an overactive amygdala, a suppressed hippocampus, and chronically high stress hormone levels are all consistent with depression.

Even so, while the hippocampus is certainly an interesting way station in this treasure hunt, we have no reason to consider it our main target. Rather, the main target should logically be the cerebral cortex's prefrontal lobes. Certainly these are a prime counterweight to the amygdala in anxiety; and since anxiety and depression seem to follow similar paths, there is reason to suspect that the prefrontal lobes play a big role in depression.

This suspicion is strongly supported by recent evidence. For we know the amygdala has powerful ties to certain parts of the prefrontal cortex, and we have seen those ties at work. Specifically, the amygdala is strongly connected to two prefrontal areas, the orbital prefrontal cortex and the radial prefrontal cortex. Imaging studies have shown that the brains of people with major depression and bipolar disorder exhibit abnormal levels of activity in both the amygdala and these two prefrontal areas. So clearly, we can relate depression to this complex of activities involving the amygdala, stress hormones, the hippocampus, and these prefrontal areas.

That's all well and good for depression, but what about bipolar disorder? What can possibly cause that? Why do its victims pass back and forth between the two opposite states of mania and depression? And how might this relate to what we already know about depression?

If we look for answers by examining the brain, we come upon a striking fact: The left and right prefrontal lobes are not emotionally balanced. One reason we know this is because a stroke in the left hemisphere's prefrontal area sometimes leads to depression, whereas a stroke in the right hemisphere's prefrontal area tends to go the other way, toward mania. In general, the right prefrontal area seems associated with negative, downbeat, and potentially depressed feelings; while the left prefrontal area seems inclined toward positive, upbeat, and potentially manic feelings. This imbalance could help to explain why major depression strikes nearly twice as many women as men, because in women the right hemisphere tends to be more powerful and so more likely to overbalance the emotion centers of the left.

But we weren't seeking an explanation for this odd female to male ratio. We were seeking possible explanations for bipolar disorder, and it looks as though we've found one. For if the brain contained a mechanism to balance the emotional biases of the left and right prefrontal lobes, and if that mechanism were defective in some people, so that it alternately favored the emotional bias of first one side and then the other, this could account for our bipolar victims' periodic bouts of mania and depression.

Wayne Drevets and colleagues at the Universities of Washington and Pittsburgh have been investigating this possibility. They have looked for genetic defects, because bipolar disorder tends to be inherited. And they have found through brain imaging that depressed/bipolar subjects with a family history of these disorders have far less tissue than normal in a part of the prefrontal lobes called the subgenual prefrontal cortex. This area, just behind the bridge of the nose, is smaller than usual because it has a problem— a marked shortage of the glial cells vital to its work. This is not the sort of defect likely to be caused by lack of use or poorly coordinated communication with other parts of the brain. Rather, it looks like an inherited defect specific to this little region that could make it malfunction—quite possibly weakening whatever governs the emotional balance of the prefrontal lobes in a way that predisposes the brain to the back and forth emotional swings of bipolar disorder.

Besides having this mechanism that could account for bipolar disorder, we now have two general explanations for depression. We can see depression emerging if the right prefrontal lobe overbalances the left. And we can also see depression arising from events starting with the amygdala that come to involve stress hormones, the hippocampus, and the prefrontal lobes. Right now, it looks as though both of these explanations are correct. It looks as though grief or prolonged stress can prompt depression by causing the amygdala to overbalance the left prefrontal lobe. It also appears that a prefrontal lobe imbalance weakening the left lobe can produce depression on its own. And finally, it looks as though these things can work together— so that someone with an overbalanced or weakened left prefrontal lobe will be especially prone to depression in the wake of grief or prolonged stress.

We can use this understanding to dispel some of the thorniest mysteries about depression, mysteries relating to medicines and sleep.

Let's take sleep first. The direct connection between depression and REM sleep is quite dramatic. We know that drugs promoting REM sleep can worsen depression. What's more, depressives tend to enter REM sleep unusually quickly after falling asleep, and compared to healthy people their REM sleep is unusually long and strong. They also spend markedly less time than normal in the deep energy-restoring sleep stages (III and IV). Small wonder, then, that depressives often find their sleep unsatisfying; because while they are getting excess REM, they aren't getting enough of the restorative deep sleep they also need.

But the most striking thing is that banning REM sleep can counteract depression. That is, if we set up a monitoring system to awaken a depressed patient whenever REM sleep starts, we can temporarily lift the depression. So there seems to be some kind of fundamental connection here.

We suspect that the main chemical culprit of this drama is acetylcholine. Normal REM sleep goes hand in hand with high acetylcholine levels, so excess acetylcholine seems likely to bear the blame for depression's overdose of REM sleep. Recalling the dance of dreams from Chapter 6, we know that a sufficiently powerful neuromodulator imbalance favoring acetylcholine over serotonin and norepinephrine can produce psychosis in the waking state—the sort of psychosis that certain highly depressed people do in fact experience. And since the serotonin and norepinephrine suppressed by this state of affairs tend to favor reason (the front part of the cortex) over mood, while acetylcholine favors mood, it is easy to see how the imbalance would tend to allow mood (in this case inappropriate sadness) and the amygdala to dominate the scene.

We can also see how banning REM sleep can relieve depression. To begin with, ending REM should end the nighttime dominance of acetylcholine, keep more norepinephrine and serotonin circulating, and thus favor the weakened prefrontal lobes over the amygdala. Beyond that, it should cause the sleeper to get more restorative Stage III and Stage IV sleep, thus diminishing the characteristic fatigue of depression.

That still leaves us with a timing problem. Banning REM sleep quickly lifts depression, but a large array of antidepressant drugs need time to act. These drugs all appear to work by reversing the dance figure of depression—by promoting serotonin or norepinephrine or both at the expense of acetylcholine. That's essentially what banning REM sleep does. So why does banning REM sleep have an immediate effect while the drugs need two weeks or more to work?

Maybe the reason is that banning REM is pretty drastic. Among other things, it could temporarily suppress acetylcholine release while enhancing the output of norepinephrine and serotonin. We know that doing this for a short time can lift depression; but we also know that fully suppressing REM for a longer time (six weeks or so) can be fatal. So clearly, none of the antidepressant drugs is doing that. Rather, they are changing sleep architecture by modifying what happens in different sleep stages—either cutting REM sleep back to more normal patterns or else abolishing REM but changing what happens in other sleep stages (notably Stage I) so that vital REM functions can be performed elsewhere. Clearly, all this is quite a lot gentler than complete REM suppression. It may also facilitate second and third messenger effects (see the earlier discussion in Chapter 6) that take considerable time to emerge. So despite the immediate impact of banning REM, and despite the fact that the drugs work 24 hours around the clock, we can see why the drugs would need substantially more time to act.

That's especially true because success depends not just on altering neuromodulator balances or strengthening the left prefrontal area but on subduing the amygdala. To begin with, we know that the relative power of the amygdala and prefrontal lobes plays a key role in depression. Beyond that, a clear association has been found between the amygdala's metabolism (level of activity) and the severity of depression. Taking this one step further, the average patient responding positively to antidepressant drug treatment shows a declining rate of amygdala activity over the course of treatment. And finally, patients with persistently high levels of amygdala activity during the remission period run a high risk of having a relapse.

Getting even a partial handle on mysteries like these shows how far we have come in recent years. But if brain science has made such rapid strides, why can't that translate into improved treatment methods?

We think it can. For example, being able to predict the chances of relapse is no small thing, because it can single out people needing further treatment before relapse occurs. And even though the imaging techniques able to do this are expensive and not yet widely available, the very existence of this potential provides an incentive to make them more affordable and available as time passes.

Or else, true bargain hunters should consider this: The future effectiveness of an antidepressant drug may well be predicted by observing how much the drug quickly corrects a depressed patient's distorted REM sleep patterns. We know what those distortions are (shortened time to first REM period, unusually long and strong early REM periods). Anyone with a little training can detect changes in those patterns by using the "Nightcap," an inexpensive recording device that the patient can wear at home (Figure 9.1). And since those changes come within a couple of days of starting treatment, there may be no reason for waiting weeks to find that a given drug is not going to pan out.

This Nightcap was devised at the Massachusetts Mental Health Center's neurophysiology laboratory some years back as an affordable substitute for expensive sleep lab monitoring. Worn on the patient's head, it reliably monitors two things: movement of the patient's eyes and movement of the patient's head. Most eye movements are confined to REM sleep; and most head movements (like most body movements) happen as the patient is entering or leaving REM sleep, because during REM the sleeper is paralyzed, and in other sleep stages the level of brain activity is too low to generate much head or body motion. So by monitoring REM sleep in two ways, the Nightcap can reliably assess changes that may predict whether or not a given antidepressant drug will be effective. This should be of interest to anyone with a serious desire to cut health care costs and improve the reliability of drug treatment for depression.

More broadly, charting the pathways of stress, emotion, and depression within the brain is leading toward new medicines and new treatments. For instance, knowing that a neuromodulator called "substance P" was involved in handling stress—and was active in places like the hypothalamus, hippocampus, and amygdala—led researchers to think that something blocking substance P might be useful against depression. Following up on this, workers have conducted preliminary tests of an experimental substance P blocker called MK-869 and have found it to be as effective as Paxil (a popular SSRI) against depression, perhaps paving the way for a new

176

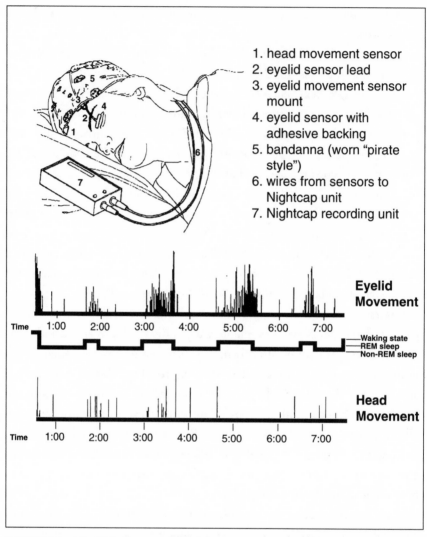

FIGURE 9.1   The Nightcap, a two-channel sleep monitor that distinguishes the states of
REM sleep, non-REM sleep, and wakefulness. One channel (the eye sensor, consisting of an ad-
hesive-backed piezoelectric film) monitors eyelid motion caused by eye movements. The other
channel (the head sensor, a cylindrical multipolar mercury switch taped to the forehead) detects
head rotations. These sensors are connected by long (3-foot) cables to the nightcap's recording
unit, which contains signal detectors, a clock, and a small battery-powered "NC Analyzer" microprocessor.

    As shown in the charts, most eye movements occur during the waking state and REM sleep,
while the statistical likelihood of head rotations is greatest when the subject is awake and during
brief periods before and after REM sleep. Therefore, by applying an appropriate algorithm, the
microprocessor can accurately assess which of these three states the subject is experiencing, as
indicated by the line appearing under the "eyelid movement" chart.

class of antidepressants that do not directly target serotonin. All of this is exciting to theorists, because it could lead to new discoveries about how depression works. But it also has strong practical implications, because not all depressed patients respond well to the currently available antidepressants, and so a novel drug with a different mechanism of action might permit effective treatment of more people.

Of course, even with this new substance (assuming it proves effective) we would still be doing what we really want to avoid—bathing the whole brain in drugs that change its chemical nature. But as we learn more about the brain structures and pathways of the mood disorders, we begin to see that treating one affected structure or one malfunctioning way station in the circle running from the deep structures to the prefrontal lobes and back stands some chance of fixing the disorder. For instance, a new technique called magnetic stimulation exposes the left cerebral cortex of a depressed patient to powerful magnetic fields. This method, which is still experimental but understandably more acceptable to patients and the public than shock therapy, provides some patients with significant temporary relief and might therefore eventually serve as an alternative to shock therapy. In this same vein, we know Parkinson's disease can be treated effectively by planting a small electrode (something like a pacemaker) inside the thalamus to break up abnormal slow wave patterns. And since at least some depressives exhibit similar slow wave patterns in slightly different areas of the thalamus, it seems reasonable to ask whether some sort of similar localized approach could yield desirable results.

Finally, we know that people suffering from long-term grief, major depression, or mania typically recover on their own. We aren't sure how that works. But we now have a good enough view of Doctor Doom and the mechanics of mood disorders to probe this question. And finding out how the brain resolves mood disorders naturally could provide strong clues to better treatment.

So the search goes on. Right now, brain science has no perfect answers. But clearly, we have learned a lot. Indeed, we have learned so much that knowing how the brain works has become vital to knowing how best to treat the mood disorders—irrespective of whether one is using psychotherapy or drug therapy or both. For assorted social, legal, and other reasons, most people now afflicted with mood disorders, even life-threatening ones, fail to get proper treatment. That's the fault of society, not science. But if we want to break out of this sad state of affairs, we will need to ensure that our practitioners have a reasonable understanding of how mood disorders actually work. And that means they should obtain enough of a grounding in brain science—either on their own or through the neurodynamic approach described later—to keep abreast of brain science advances now and in the future.

10

The Land of Voices

Psychiatrist: How are you today, Nick?

Nick: (silence)

P: Did you enjoy your breakfast?

N: (after a long pause) I didn't understand. . .

P: Your breakfast. Do you remember, you weren't enjoying your food?

N: (silence)

P: What did they give you for breakfast today?

N: Please would you ask me again. . . (long pause of three minutes). . . I really can't answer now.

—Interview with a schizophrenic patient showing poverty of speech. From Ming Tsuang and Stephen V. Faraone, *Schizophrenia, The Facts* (2nd edition), 1997

Any ill that is mysterious stands a better chance of being poorly dealt with than one that is understood. That puts schizophrenia in a good spot to be handled badly, because it is the most mysterious of all leading mental ills. Indeed, its very name sows confusion. The old name for schizophrenia, "dementia praecox," means early mental loss. While that clearly falls short of full description, the current name also lends itself to misunderstanding. For schizophrenia (literally "split mind") is commonly regarded not as splitting or fragmenting mental processes (which it does) but as splitting the victim's personality. In truth, however, schizophrenia is not a split or multiple personality disorder, or even one that necessarily makes its victim undecided. Instead it afflicts thinking, that most human of attributes, its hallmark being a profound mental disruption that makes it hard for affected people to organize thoughts and feelings. This disruption produces a mix of symptoms, ranging from hallucinations to incoherent speech, not all of which need be present in any given case or at any given time in order for the disease to be present.

Unfortunately, that's just the start of the diagnostic problem. Schizophrenia can also be confused with other ills, because many conditions (tuberculosis, epilepsy, autism, and others, as well as certain drug states) can mimic its symptoms. It often goes hand in hand with mood disorders, blurring the line separating it from these other mental ills and giving rise to a separate diagnosis of mood disorder combined with schizophrenia that is called "schizoaffective disorder." And finally, the distinction between schizophrenia and lesser ills is blurred because a lot of people, especially close relatives of schizophrenics, tend to show milder schizophrenia-like symptoms that qualify them as having something called "schizotypal personality disorder," and sometimes the line between this and schizophrenia is hard to draw.

According to DSM-IV, to be diagnosed as schizophrenic a person must have at least two of five symptoms for a "significant part" of a one-month period. These symptoms are as follows: hallucinations (usually the schizophrenic's own inner voice mistaken for imaginary voices); delusions (firmly held erroneous beliefs, commonly of being watched or followed); grossly disorganized or catatonic behavior; disorganized speech; and any of various "negative" symptoms including flattened emotion (subnormal expression of emotion), impoverished speech (where the person says little or the words have little meaning), and avolition (lack of will). The person must also have experienced difficulty with self-care, work, or interpersonal relations; must have had continuous signs of mental disturbance for at least six months; and must not have any other condition (e.g., autism, schizoaffective disorder, epilepsy, or drug abuse) that could account for the problem.

Of course, none of this tells us what schizophrenia really is. In fact, many experts still throw up their hands and say "we just don't know." But in fact we have learned a lot. We have some good ideas about causes; we are learning about relevant brain chemistry and also about coordination of brain regions responsible for directed thought and emotion; we have some powerfully effective drugs; and we have a lot of experience with therapy and related measures. So although our knowledge is incomplete, we see no reason for giving ground to this terrible scourge by pleading ignorance. Rather, it seems worth reviewing what we *do* know (as well as what it looks reasonable to suspect) and seeing where we can hope to go from there.

A good way to start getting a feel for schizophrenia is to look at an actual case history. That can give us a close look at schizophrenia's symptoms and how they develop. The case we have selected* also provides a feel for the ups and downs of the disease; the kinds of problems that a victim can run into; what happens when a patient is well supported and effectively treated, as opposed to unsupported and untreated; how both the classic and the newer "atypical" antipsychotic medications work; and why sound monitoring, social support, and follow-up are needed.

Alfred Ramsey, a 41-year-old bachelor of mixed African, European, and Cherokee Indian descent, was first seen at Boston's Massachusetts Mental Health Center in 1995. However, his troubles had actually started some twenty-two years earlier, at age 18, when he was attending college and living at home with his parents. In those days, according to his parents, he began behaving strangely—whispering to himself, removing the mirror and television from his room without saying why, and regularly inspecting the furniture as if looking for something. One night he left the house through a window to visit a fundamentalist church in another city. When he returned, his parents urged him to seek treatment. However, he refused and remained in college for two more years, until he failed a course and was placed on academic leave.

---

*For a more detailed description of this case see S. Rafal, M. Tsuang, and W. Carpenter, Jr., *American Journal of Psychiatry*, July 1999, pp. 1086–1090.

One day, when he was 21, Alfred Ramsey raised all the shades in the house to let in light and drive out "evil spirits" that he claimed might be listening to his thoughts. All phases of his sleep were disturbed, producing severe fatigue, and he was admitted to a psychiatric hospital. There he was reported to have put his fingers in his ears when he heard the intercom, to have made bizarre faces, and to have shown marked social avoidance, sharply impaired attention, limited ability to make associations, derailed thought, and blunted feelings. He was treated with oral trifluoperazine, an antipsychotic drug, together with another drug, benztropine, to reduce side effects (stiffness and tremor) produced by the first. Two months later he left the hospital with a discharge stating that he had experienced an "acute schizophrenic episode" but had then shown some improvement.

For the next 15 years Alfred Ramsey continued living at home with his parents. He remained moderately disorganized and socially withdrawn, but persistently denied that he was ill. Because of his parents' monitoring, he generally took his medicine (consisting of one or another of the antipsychotics known as "neuroleptics" that are often prescribed for schizophrenia—including trifluoperazine, fluphenazine, haloperidol, and thioridazine—usually in combination with benztropine to suppress stiffness and tremor). During this period his parents' attention, plus periodic visits with psychiatrists versed in pharmacology and with other counselors, helped to track his condition, determine when his course of medication should be altered, and keep him in touch with his own treatment. Probably as a result, Ramsey was able to complete an associate's degree by taking courses part-time. He then worked occasionally at temporary manufacturing jobs, his last employment ending at the age of 33.

To this point the relatively benign course of Ramsey's schizophrenia is not hard to explain. To begin with, he stayed in the mental hospital for two months following his first admission. This provided adequate time to assess his illness and personal situation, prescribe appropriate medication, and affirm that the prescribed medicines were both acceptable to him and taking hold. Equally important, as noted above, from then on he was living with two actively concerned parents who could keep tabs on him, monitor his treatment, see that he took his pills, report any changes in his condition, and generally coordinate his care with a psychiatrist and other health care workers. In effect his parents, who could hardly help but become highly familiar with his disease symptoms, took the place of both community outreach services and supervised live-in personnel. Obviously, while some mentally ill people enjoy similar family assistance, many are not so lucky.

Ramsey began getting into trouble when for uncertain reasons (aging parents, resurgent illness, drug side effects, or a yen for independence) he broke through the protective shell provided by his family. At age 30, and again at 33, he failed for a time to take his prescribed medicine, had a psychotic flare-up, and was hospitalized. The record from the first of these hospitalizations refers to inappropriate emotion and reluctance to reveal his thoughts—which appeared to be blocked when they could be inferred from his speech. He acknowledged that when he stopped taking medication his concentration became poor, even though he denied that he had an illness. His discharge diagnosis was one of "chronic paranoid schizophrenia."

The slowly gathering storm broke four years later when Ramsey left his parents' home for the first time and moved into his own subsidized apartment. There he stopped taking his fluphenazine pills almost immediately, and his housekeeping fell apart. The Boston Housing Authority reported so much trash in one room that an inspector was unable to enter. Informed that this was causing a cockroach problem, Ramsey's response was to place plastic wrap and aluminum foil on the walls to keep out roaches. He risked injury by wandering the streets at all hours of the day and night and by taking in homeless strangers. On one or more occasions his mother expressed alarm at his condition—which appears to have included poor hygiene, poor nutrition, disorganized thoughts, and paranoid delusions.

Probably because of the laws against involuntary confinement, it took over a year for Ramsey, then 39, to reenter a psychiatric institution. The record indicates that by then he was hearing voices, speaking little, and exhibiting disorganized thoughts, blunted emotions, and social withdrawal. The institution presented a legal petition for full guardianship, and at Ramsey's request a court-appointed attorney rather than a family member was made his guardian. Largely because he had developed undesirable side effects to the standard antipsychotic neuroleptic drugs, he was prescribed a new drug (perphenazine), together with benztropine for the lingering side effects of earlier medication. He was later transferred to a day hospital program located at the Massachusetts Mental Health Center. The program and its associated shelter were geared to provide continuous care—including biweekly pharmacologic monitoring, supportive psychotherapy, case management, supervised living, and effective follow-up.

The next week he left the program, returned to his old apartment, and stopped taking his pills. That looked like the start of another downhill plunge. However, the plunge was cut short ten days later when he was hospitalized involuntarily—by enforcing the guardianship agreement. When Ramsey said he needed medicine for his "jumbled thoughts," it was the first time on record that he had asked for medication to help his thinking. He then returned to the Mass Mental program, where he remained highly concerned about what he termed his "independence" and the need to return to his apartment. The main problem, however, was that he had developed adverse side effects to his old antipsychotic drugs, and his new drug was not very effective against his ailment.

At first he was continued on the same drug regimen, but his condition remained poor. He rarely spoke spontaneously, and what he said was usually brief, vague, and disorganized. Apart from wanting his own apartment, he showed no signs of having specific interests or desires. He made few if any social contacts, sought little activity, and was inattentive at the group meetings that were part of his treatment. He also showed little insight into his illness and remained paranoid, explaining that he could not live in a group home because "I can't trust anybody there."

Then came a dramatic turn. Over time, the typical neuroleptics commonly used to combat psychosis had caused Ramsey significant stiffness and tremor, to a point where these side effects were disabling—despite sizable doses of benztropine prescribed to suppress them. So instead of returning to such drugs, Ramsey's Mass Mental doctors decided to try a new antipsychotic called clozapine that reports indicated was less likely to produce these disabling effects and that had a broader range of neuromodulatory action. Not all schizophrenia cases respond well to clozapine; and because clozapine can occasionally cause a fatal blood disorder called agranulocytosis, weekly blood tests would be needed. Despite these negatives, however, the prospects for a worthwhile result seemed good; and so the decision was made to go ahead.

At first the new drug produced mainly sleepiness and excess salivation, but in about six weeks Ramsey's schizophrenic symptoms began to ebb. His feelings brightened; his level of interest rose; and eventually (some four months later) he said his thoughts were clearer. Indeed, his thoughts did seem better focused. For the first time he began participating, without prompting, in the daily hospital group meetings; he paid attention to the discussions; and he astonished staff members by offering support and understanding to other patients. These dramatic changes were all the more remarkable because they occurred after twenty years of essentially continuous disability.

There were other milestones too. He began attending a psychosocial "clubhouse," where (with encouragement) he helped with cooking, cleaning up, and assorted community endeavors. He acknowledged that he had a mental illness called "schizophrenia"; and while fearing the effects of the weekly blood tests and voicing misgivings about clozapine, he continued faithfully taking the prescribed pills on his own—as demonstrated by the levels of clozapine found when his blood was tested. He began hoping that he might return to work someday, even that he might find a girlfriend. And after spending sixteen months in a psychiatric shelter, resisting suggestions that he move to a group home, in July 1996 he returned to his own apartment, which he proceeded to maintain effectively.

The story does not end there, however. Despite the drug's beneficial effects, Ramsey still lacked confidence in clozapine, remained unduly fearful about the long-term impact of weekly blood tests on his health, and expressed interest in switching to some alternative drug if any came along. For these reasons, and also because clozapine caused Ramsey considerable daytime sleepiness, when the antipsychotic drug olanzapine was released for general use in October 1996, his doctors suggested switching him to that.

This was not a snap decision. The doctors knew the reasons for not changing drugs. They were well aware of the substantial progress Alfred Ramsey had made with clozapine—progress that would be risked by switching to the new and relatively untried olanzapine. They knew his theory that blood tests were doing cumulative damage to his health were false. They knew he had a history of poor insight into his condition. They had no assurance he would be willing to switch back to clozapine if olanzapine produced a poor response. They recognized, despite the guardianship agreement, that "compliance with taking oral medication is ultimately (and literally) in the patient's hands." And they knew that some patients who had left clozapine and then gone back to it had responded less successfully the second time.

So why did they go ahead? There were several reasons. For one, long-term adherence to any treatment regimen depends on physician–patient collaboration. So Ramsey's long-term cooperation was likely to be enhanced if he felt he was a respected partner with power to influence his treatment plan—especially since he had shown a strong desire to move toward independence, a goal his doctors felt they should make every reasonable effort to support. Beyond that, olanzapine appeared to be a reasonable pharmacologic alternative to clozapine. Its affinity for particular neuron docking sites seemed reasonably similar to clozapine's; but it had less of an antihistamine effect, so there was reason to hope it would be as effective as clozapine without causing sleepiness. And finally, schizophrenia patients' attitudes toward treatment have an impact on their long-term response to treatment—so Ramsey's positive attitude toward olanzapine and negative attitude toward clozapine seemed likely to influence the effectiveness of treatment.

The doctors discussed the risks of making the switch with Ramsey and his guardian, and received their consent to go ahead. Then, proceeding gingerly, they very slowly reduced Ramsey's daily dose of clozapine and added small comparable amounts of olanzapine. Finally, three months later, Ramsey took his last dose of clozapine and from then on depended entirely on the new drug.

This change did not notably worsen any of Ramsey's schizophrenia symptoms. Instead, significant improvements occurred. His sleepiness abated; he became more attentive and less likely to lose his train of thought. As a result, he became more actively interested in his psychosocial clubhouse, attending some three meetings a week. Indeed, he came to play a central role in a wide range of clubhouse community activities. And before the available published record ends, in early 1999, he had began attending church; he was visiting his family more often; and he had begun to talk in concrete terms about finding a part-time job.

In the end, the doctors treating Ramsey found his warmth, engaging manner, and relatively well-preserved social graces to be unusual for a patient with a twenty-year history of schizophrenia. That's because most long-term schizophrenics show far more deterioration of their interpersonal skills. However, they also felt that cases like Ramsey's might well become more common, given the increasing use of clozapine, olanzapine, and other "atypical" neuroleptic drugs. For besides blunting schizophrenia's "positive" symptoms (hallucinations, delusions, grossly disorganized or catatonic behavior, and disorganized speech), these drugs often have salutary effects upon problems relating to cognition (perception, thought, and memory). They may also reduce negative symptoms, though how much impact they actually have on primary negative symptoms (such as flattened emotion, impoverished speech, and lack of will) is uncertain.

Within the context of unusually successful treatment, this account illustrates several important points about schizophrenia, namely:

- The disease typically strikes in late adolescence or early adulthood, with most victims being over 20. (While it can occasionally strike children and people over 40, such cases are relatively rare.) It thus wreaks havoc in what should be the prime of the victim's educational, vocational, and sexual life.
- Schizophrenia typically waxes and wanes. This spontaneous fluctuation, combined with the symptomatic relief that drugs provide, can result in long periods of relative health. But the risk of relapse

remains—as most patients who experiment with medication "holidays" discover.

- The disorder is tenacious but not hopeless. Before 1955, when Thorazine was introduced, many patients spent their lives in mental hospitals. By 1970, when many of the hospitals had closed, the medication options had increased. Now, at the dawn of the new millennium, even the most difficult patients have some chance of living a relatively normal life.

What has brain science done to flesh out this general picture? The answer is quite a lot. Among other things, clinical studies have taught us a good deal about the genetic and biological roots of schizophrenia. A wide range of experiments have begun to show the relevant brain chemistry involved. And brain imaging backed by other research has revealed a lot about how schizophrenia affects brain structures and their work.

To begin with, a host of studies have sought to separate family, social, biological, and other influences from genetic factors. These have revealed that schizophrenia has strong genetic roots. For this reason, relatives of schizophrenics are more likely than other people to develop the disease. On the average, compared to the general population's roughly 1 percent chance of getting schizophrenia, first cousins of schizophrenics have about a 2 percent chance, nieces or nephews have a 4 percent chance, a fraternal twin has a 17 percent chance, and an identical twin has a 48 percent chance. Of course, if genetics told the whole story then all identical twins of schizophrenics would get the illness. So we also need to look at nongenetic factors.

Despite the Freudians' old focus on "schizophrenogenic mothers," however, a wealth of studies has provided no compelling evidence that either maternal or any other family behaviors cause schizophrenia. Nor does low socioeconomic status appear to cause schizophrenia: because schizophrenics tend to be jobless, schizophrenia tends to cause low socioeconomic status; but the reverse is not true. Indeed, no family or social factors seem to play a large role in causing schizophrenia; and so investigators have increasingly focused on biological factors, the most likely of which are pregnancy and delivery complications and to a lesser extent viral infections.

Various studies have shown an unusually high rate of pregnancy and delivery complications among children who ultimately become schizophrenic. That's not too surprising, because both pregnancy and delivery problems (as well as viral infections) can cause subtle as well as gross brain damage. But while this rate of pregnancy and delivery complications is somewhat higher among schizophrenics than among the general population, the difference is not dramatic.

What is dramatic are the findings of studies combining pregnancy and delivery complications with genetic data. These studies suggest that delivery complications have a marked tendency to trigger schizophrenia among children who are already genetically inclined in that direction. In such cases the defective genes appear to place the explosive charge, as it were, while delivery complications light the fuse. This suggests that the reason schizophrenia strikes one genetically predisposed identical twin but not the other could relate to delivery complications or other factors that affect one genetically predisposed twin but not the other.

This suggestion that delivery complications and other factors may merely be activating a genetic predisposition makes genetics and the 48 percent identical twin figure loom large. But study after study has failed to pin the tail on the donkey. That is, try as we might, we have been unable to single out the guilty genes. This doesn't mean that genes are not involved, and in fact we have fingered a number that look suspicious. But it does imply that no single gene or easily identifiable group of genes causes the disease. Instead, it seems reasonable to suppose that a diffuse horde of genes does the job, with a different collection of genes acting in nearly every case—which would make it easy to understand why no individual genes stand out and also why the disease's severity and symptoms vary so.

Regarding brain chemistry, we have known for some time that the classic antipsychotics, which first became available around midcentury, act by blocking dopamine receptors. Specifically, they block what is called the D_2 receptor, one of five known dopamine docking sites. These drugs (chlorpromazine, haloperidol, fluphenazine, and others) all tend to have similar effects. That is, they tend to reduce the "positive" symptoms of schizophrenia and combat psychosis, regardless of whether the psychosis arises from schizophrenia or something else. They produce desirable responses in about 70 percent of the schizophrenics who take them. Like the antidepressives, they need time to act, typically taking effect over a five- to ten-week period. They do not relieve the negative (passive) symptoms of schizophrenia. And presumably because they affect a kind of dopamine receptor highly concentrated in the caudate nucleus and other areas involved with motion, they tend to cause side effects termed "extrapyramidal" that include tardive dyskinesia and that are reminiscent of the stiffness and tremor of Parkinson's disease—a brain ailment associated with insufficient dopamine.

That doesn't mean that schizophrenia is caused by excess dopamine. Rather, schizophrenics seem unusually sensitive to dopamine. As we have seen, amphetamines and cocaine can produce schizophrenia-like psychosis, and schizophrenics respond unusually strongly to these drugs. Also, experiments with animals showing schizophrenia-like symptoms have pointed to abnormal dopamine sensitivity. In addition, schizophrenics show some evidence of having excess D_2 receptors in parts of their deep cerebral structures that might make them unusually sensitive to dopamine.

So we really don't know whether dopamine levels are being altered or whether the changes we observe are being wrought exclusively by increased neuron sensitivity to dopamine. In either case, we have no evidence that the genetic or other factors producing schizophrenia are causing these dopamine-related changes directly; so it seems likely that these changes are just one link in a chain of causation—a link vulnerable to remedial action by our drugs.

This impression gets reinforced if we look at "atypical" antipsychotics like clozapine, risperidone, olanzapine, quetiapine, and others that have become available more recently. Some of these drugs block the D_1, D_2, and D_4 dopamine receptors to one degree or another; but different ones have an even broader spectrum of neuromodulatory effects—blocking assorted acetylcholine, serotonin, norepinephrine, and histamine receptors. They influence schizophrenia differently than the older drugs, and they work in different ways. For example, clozapine proves effective for roughly 30 to 50 percent of all schizophrenics who fail to respond to the classic antipsychotics; most of the atypicals cause relatively few Parkinsonian side effects (a good indication that they may also cause less slow-developing tardive dyskinesia); and some work against anxiety disorders and depression as well as against schizophrenia, providing another pharmaceutical means of treating those problems.

Unlike both the older antipsychotics and the other atypicals, clozapine can cause the potentially fatal blood disorder mentioned earlier in about 1 percent of the cases, and so it requires careful monitoring. But as the Ramsey case shows, clozapine as well as some of the other atypicals sometimes appear to work against the negative symptoms of schizophrenia as well as the positive ones, suggesting another major gain.

All in all, it seems clear that the new atypicals have opened a new path for treatment of mental ills. The natural urge of any drugmaker is to develop a "clean" product, one that will have a specific impact at one site. But maybe that's not best for treating mental ills. Maybe hitting one neuromodulator at one kind of docking site is more likely to unbalance the brain and cause adverse side effects. Maybe a drug that has the same impact but does its work more gently through multiple modulators at several different sites can do a better job. In any case, the atypicals have shown that drugs that operate this way can work effectively; and since the atypicals, unlike the classic antipsychotics, are quite different from one other, they are clearly inviting further exploration of this promisingly broad path with its wide diversity and many points of action.

Our resulting embarrassment of interpretive riches has made it hard to pin down specific therapeutic mechanisms. But one thing seems evident. That is, the psychosis of schizophrenia, like all other psychoses, involves more than just dopamine at the D_2 docking site. For the dancing neuromodulators interact both with each other and with neurotransmitters like glutamate and GABA; and many kinds of docking sites are involved, each of them distributed differently within the brain.

While serious efforts have been made to track schizophrenia through these neuromodulator/neurotransmitter thickets, most have had very limited success. Nor have we succeeded in tracking schizophrenia to its lair in some particular brain structure. We know the prefrontal lobes are involved. Among other things, they are the headquarters of the thought and judgment that schizophrenia clearly harms; and we know that damaging a monkey's prefrontal cortex before birth in certain ways can lead to behavior problems surfacing in adolescence that bear some resemblance to schizophrenia. But we also think that the temporal lobes, in charge of speech and language, are involved. That's because imaging studies of schizophrenics strongly suggest a breakdown in normal communication between the frontal and temporal lobes, a monitoring failure that apparently makes the victim attribute his own inner speech to another person, thereby giving rise to auditory hallucinations. Beyond that, we need to cast our nets even wider, because schizophrenia can cause other symptoms—including motor problems (catatonia), emotional difficulties (affective blunting), and other troubles—that do not seem to be centered in the prefrontal lobes, or even in the frontal and temporal lobes. So increasingly, brain scientists have come to view schizophrenia as caused by weak or defective coordination affecting many brain areas, and they have sought to develop models reflecting this.

One such model, proposed by noted brain scientist Patricia Goldman-Rakic, relates to working memory. Working memory, centered in parts of the prefrontal cortex, enables the brain to hold a concept or representation "on-line" and perform operations with it. In our dreams we have all experienced the impairment of working memory that is caused by the normal inactivation of the prefrontal cortex areas in REM sleep. We are disoriented and confused; we hallucinate, confabulate, and forget.

Defective working memory could account for various schizophrenia symptoms. For example, failure to effectively monitor spoken and unspoken speech output could lead to disorganized speech and also to auditory hallucinations. Failure to plan behavior could lead to negative symptoms such as lack of will and impoverished speech. And failure to match a specific experience with associative memories stored and coordinated elsewhere in the brain could promote delusions. The model suggests that certain prefrontal regions—and also their links to other parts of the cerebral cortex, the thalamus, and the corpus striatum—could play a major role in the disease.

Another model arises from schizophrenics' common complaint that their senses give them more information than they can handle. It suggests that this overloading causes both their misinterpretations (delusions and hallucinations) and their "retreat to safety" (negative symptoms). To test this model, researchers have worked with the "startle" response triggered by a sudden bright light or loud noise, a response shared by humans and all mammals. In normal subjects, the startle response is generally reduced if a weak stimulus occurs shortly (some thousandths of a second) before the bright light or loud noise. But in schizophrenics the reduction is less than normal.

Rat experiments have shown that brain circuits passing through the cerebral cortex, corpus striatum, and thalamus play a key role in this process of reducing the startle response. They also show that reduction of the response can be eliminated by enhancing dopamine action, and that the reduction can be restored by both classic and atypical antipsychotics. Overall, this work points up the likely importance of these cortico–striatal–thalamic brain circuits in schizophrenia.

A third model, backed by an impressive body of accumulated evidence, blames schizophrenia on poor coordination by brain regions including the prefrontal cortex, the thalamus, and the cerebellum. According to Nancy Andreasen and others who devised this model, disruption in this circuitry makes it hard to prioritize, process, coordinate, and respond to information. This difficulty produces a basic mental deficit that can account for schizophrenia's wide array of symptoms.

Of course, Andreasen's team is not alone in implicating the prefrontal cortex. Just about everyone doing schizophrenia research these days sees the prefrontal lobes as playing some kind of role. But naming the thalamus and cerebellum as major actors is less usual, so it seems worth asking why they are singled out.

Let's start with the cerebellum. It used to be thought that the cerebellum's sole job was to coordinate motion. That's a big job all right, but even so we might have thought it was doing more, because the cerebellum's recent evolution resembles a bull market romp on Wall Street. Indeed, it matches the rapid evolutionary development of the prefrontal lobes, because both the cerebellum and the prefrontal lobes are proportionately one-third larger in the human brain than in the brains of nonhuman primates. The cerebellum is also more deeply furrowed than the cerebrum; it contains more neurons than all the rest of the brain combined; and it connects to the rest of the brain with neuron cables able to transmit roughly forty times the information carried from the eyes to the brain by the mighty optic nerve.

So, what is the cerebellum actually doing? As researchers discovered in the 1990s, the cerebellum coordinates a lot more than motion. In fact it appears to coordinate memory, vision, spatial orientation, touch, language processing, planning, foresight, judgment, attention, motivation, emotion, establishment of conditioned responses, integration of higher-order behavior, and probably other things. Naturally, the cerebellum is not the sole or lead actor performing any of these tasks. But it receives a vast torrent of information from parts of the brain dealing with all these things; it has a very regular structure with tremendous computing capacity; and so it appears to act as the brain's computer, blending and harmonizing many kinds of information much as a skilled orchestra conductor blends and harmonizes the work of numerous musicians. According to neurologist Jeremy Schmahmann, a pioneer in cerebellum research at the Harvard Medical School, "the aim [of the cerebellum] is to take information, smooth it out, and make it harmonious with the intended goal—regardless of whether this is a motor goal or some other goal. This implies that when you lose the cerebellum's contribution to a given task, the task may still be performed, but will be performed poorly."[1]

To assess the clinical implications of this discovery, Schmahmann conducted a seven-year study of twenty people with brain damage limited to the cerebellum. That study showed a clear pattern of behavior changes. Typically, the patients exhibited reduced ability to plan, reason, make decisions, initiate actions, and change strategies; trouble with spatial organization, grammar, and language; reduced IQ; impaired working memory; and very flat emotions. Not surprisingly, considering the cerebellum's involvement with so much that the brain does, these impairments involved many functions commonly ascribed to the cerebral cortex and other cerebral structures.

Many of these problems are also mildly reminiscent of ones found in schizophrenia. That's not surprising either, for we know the cerebellum has strong connections to the prefrontal lobes as well as to the thalamus, and several studies have shown various prefrontal and cerebellar activities to be related. Beyond that, a study of fourteen elderly schizophrenics found that large neurons in their cerebellums called Purkinje cells had a cross-sectional area 8 percent smaller than Purkinje cells in thirteen elderly control subjects. Other studies have shown increased blood flow in the cerebellum when schizophrenic patients are taken off antipsychotic medication. And yet another study has associated schizophrenia with failure to activate portions of the above-mentioned circuits connecting the cerebral cortex, cerebellum, and thalamus.

All of which brings us to the thalamus. As previously noted, the thalamus does the job of a gigantic relay station, but it is far more than a relay station. It receives, filters, and forwards nearly all incoming sensory information. It receives and reroutes the output from the cerebellum. It has strong ties to the prefrontal lobes. It coordinates myriad neuron "loops" running out to many parts of the cerebral cortex and returning. It also communicates with many other areas, including the emotion and memory centers deep in the cerebrum. And of course, as seen in Chapter 6, it appears responsible for coordinating consciousness. That means that the thalamus relates intimately to a wide range of conscious activities; and because schizophrenia also does this, it should come as no surprise to find the thalamus deeply involved with schizophrenia.

Evidence of such involvement has shown up in several ways. A 1990 study found that, compared to normal subjects, schizophrenics had substantially fewer neurons in a part of the thalamus called the medial dorsal nucleus—this being the main thalamic nucleus communicating with the prefrontal cortex. Also, each of four magnetic resonance studies conducted by the Andreasen team found the average size of the thalamus to be smaller in schizophrenics than in normal subjects. Beyond that, a 1995 positron-emission tomography (PET) study found that schizophrenics who were experiencing auditory hallucinations showed abnormal activity in the thalamus. And another PET study found that the front left portion of the thalamus was smaller than normal in twenty schizophrenic patients. All of these findings support the idea that the normal role of the thalamus is disrupted in schizophrenia. Of course, none of this precludes involvement of other brain areas—among them the corpus striatum, which has been implicated by other studies and which also has close ties to the thalamus.

All this makes the Andreasen model compelling. But not everyone agrees with it, even though most of today's theorists seem to agree on certain basic points. That is, most experts now think that schizophrenia's symptoms arise from disruptions in certain basic brain circuits, and that its symptoms can be reduced by taking medicines that affect those circuits. Most link this disruption to poor coordination of experience with incoming information as the schizophrenic tries to cope with the real world. Most think the disease involves widely distributed brain circuits rather than any single locale within the brain. And most believe that relationships between the prefrontal cortex, other interconnected parts of the cortex, and various subcortical regions play a key role in the disease.

As this suggests, we lack general agreement on anything but a very muddy picture. For besides being general, these points of agreement fail to pin down schizophrenia's basic cause, why its features vary so from case to case, why it takes so long to appear, and how it relates to brain chemistry. All these matters are important. But we will touch on the first three a little later in this chapter, so the main question to be dealt with here is how schizophrenia changes brain chemistry.

One astonishing possibility is that it doesn't—at least not dramatically. We have some indirect evidence of chemical changes in the brains of schizophrenics and their relatives; but we have no clear indication that schizophrenia greatly alters prevailing levels of serotonin, norepinephrine, acetylcholine, or any other neuromodulator. The disease clearly involves dopamine. But the brain handles dopamine gingerly, keeping dopamine levels fairly constant throughout its various sleep and waking states, and we have no compelling evidence that schizophrenia alters that. Instead, as noted earlier, the brain of a schizophrenic may simply increase certain neurons' sensitivity to dopamine.

If so, the lack of a dramatic dopamine shift would fit well with the idea that schizophrenia arises fundamentally from structural problems afflicting thought. For in that case it would seem a poor idea to change neuromodulator levels. Raising acetylcholine would push things the wrong way (favoring emotions over thought); raising serotonin or norepinephrine might enhance thought but could also do undesirable things like alter sleep architecture; and raising dopamine would influence thought but would also affect the other dancing neuromodulators, would act throughout the brain, and would seem more likely than raising selective neuron sensitivity to bring on psychosis—something that the brain is presumably trying to avoid.

In contrast, a selective increase in the dopamine sensitivity of certain neurons or groups of neurons might be directed at improving precisely those thought processes that need fixing and would seem a better bet for accomplishing desired alterations. Even so, the common emergence and persistence of psychosis suggests that any such adjustment mechanism may be fairly crude—like raising the body's temperature (causing fever) to fight off germs, or raising the gain on a ham radio set to improve reception. Also, there is a balancing problem. If the brain hikes dopamine sensitivity too little it will have trouble processing its thoughts (yielding to the negative symptoms of schizophrenia); but too much of a rise can cause imaginary thought patterns (voices, delusions, and other positive symptoms) leading to psychosis; and there seems no guarantee of finding any reasonable middle ground between.

This theoretical picture provides a new possible slant on why our psychiatric drugs are helpful. For despite their drawbacks, the drugs can do things that a selective dopamine adjustment system probably can't do. Notably, they can home in on certain specific dopamine docking sites like the D_1, D_2, and D_4 sites; and they can also influence specific nondopamine docking sites affected by other neuromodulators, thereby engaging these other sites in the adjustment process. These effects are rather crude. But if this theory turns out to be correct, it will provide us with the comforting assurance that we are not trying to deal with schizophrenia on our own, any more than doctors armed with antibiotics are trying to deal with bacterial infections on their own. Rather, in the best traditions of modern medicine, we are trying to use drugs, therapy, and other measures to assist at least one natural adjustment process (there may be others) that the human brain has evolved to do this job.

Of course, we still have far to go. But the very fact that we can devise detailed speculative theories of this sort suggests that brain science may be getting a handle on what schizophrenia is, how it works, and how our antipsychotic drugs work against it. In any case, it seems clear that we have a far better grasp of all these things—theoretical and otherwise—than we did before brain science came along.

But if brain science has made progress, what of it? What are the practical implications?

That's an easy question to answer, because some of the practical implications are pretty clear. For instance, we no longer need experience to tell us that we cannot take a schizophrenic, lay him on a couch, probe him for repressed memories, and expect a cure. Nor can we expect any other nonmedical therapy to succeed on its own hook, because we are confronting not only psychosis but also major imbalances within the brain that defy effective treatment with therapy alone.

So we must use drugs. But what we are learning makes us edgy about that. For it looks as though we cannot expect any drug to serve as a "magic bullet." Instead we must regard our drugs as marginal answers to a lifelong ailment that is quirky and needs regular monitoring. Thus, even though our drugs have gotten better and more versatile, and even though we may be able to improve them further, we can no longer evade responsibility by pretending that drugs alone are a sound answer.

Nor should we assume that quick drug prescription based on imperfect understanding of the patient will yield desirable results. Rather, especially since we now have a wide range of different drugs available, the drugs must be carefully tailored to the patient; the patient must be confined (if psychotic) or otherwise monitored long enough to ensure that the medicine is taking hold; and treatment must be supplemented with appropriate therapy and also with a range of social services designed to reorient the patient and provide effective long-term supervision. As we saw in the case of Alfred Ramsey, such combined treatment can beat schizophrenia back a long way, not in all cases but in many. But it is unlikely to do this unless those who devise the treatment have a good grasp of what is needed, a grasp based on the still limited but growing knowledge of schizophrenia that is being gained through brain science. We will have more to say about this in Part Four.

Before leaving the brain science of schizophrenia, however, we wish to examine three remaining mysteries: What is schizophrenia's fundamental cause? Why do its features vary so from case to case? And why does it wait so long to strike? We have no firm answers here, but we have a lot of partial answers—so many, in fact, that it is tempting to try to put some of them together.

A good place to start is with genetics. Genes play a big role in schizophrenia. But the responsible genes are so broadly dispersed and obscure that concerted work by many experts has failed to identify any gene as being definitely tied to the disease. Maybe that means something. Perhaps this lack of findings suggests a connection between schizophrenia and brain evolution. For in fact a diffuse array of mildly defective genes is precisely what we would rightly expect to find trailing in the wake of rapid human brain evolution, much as cosmologists rightly expected to find low-level microwave radiation lingering in the background of a universe created by a cataclysmic "big bang."

We can see this better if we look at more humble processes like creating new cars or computers. After plans for a new car or computer are made, drawings are created, and a few prototypes are built, the last major step before production is getting out the bugs. As anyone involved can attest, there are always lots of bugs; and as the more obvious ones are removed, what remains is a diffuse array of marginal bugs that are less clearly tied to bad effects and hard to find.

Of course, brain evolution is complex enough to make car and computer building look like child's play. So if a new part of the brain were growing and evolving fast, this would seem certain to create innumerable genetic bugs—ones relating to the new part's work and also to its multitudinous connections and complex coordination with other parts.

How would these bugs be removed? Nature's tool for removing genetic defects is natural selection. So the first to go would be major defects that kill their owners when they are young, before they have a chance to reproduce. And the last to linger would be minor bugs that have to team up with a motley assortment of other bugs to do any real harm and that don't even express themselves until their owners have had plenty of chance to reproduce. That's precisely where the genes that cause schizophrenia are today.

Let's take this one step further. Since about 2 million years ago, when the brains of prehumans were literally pint-sized (with a capacity of roughly half a liter), the size of the prehuman and human brain has been growing at an average rate of about 1 cubic centimeter per 2000 years. That's a snail's pace by the standards of recorded history, but it's meteoric in evolutionary terms. And what's more, this fast growth seems to have favored certain structures—because our prefrontal lobes and cerebellum are much larger relative to the rest of the brain than they are in primates, and we can assume that the temporal lobe of the cerebral cortex has undergone major changes relating to development of language. So if a host of genetic brain defects spun out by this evolution were causing schizophrenia, we would expect to find schizophrenia tied to the fastest-evolving and therefore most bug-prone regions of the brain, these being the prefrontal lobes, the temporal lobe, the cerebellum, and their coordinating circuits in the thalamus. And in fact that is pretty much what we do find.

But we should sound one cautionary note. With respect to the natural weeding out of bugs, we are not talking about quick decisive trends. Rather, we are talking about slow trends extending over thousands of millennia of human and prehuman evolution. We know the genes involved in schizophrenia today are carried not only by schizophrenics but also by their non-schizophrenic relatives. So we would expect such genes to be weeded out quite slowly, with those most likely to cause the disease and those tending to interfere more with reproduction being weeded out faster than the rest.

Of course, such weeding would have been more or less balanced in the past by creation of new bugs as the human brain continued to evolve. So there is no reason to suppose that these selection trends would have brought about any significant decline in the disease. But there is reason to think that these trends would have discriminated against genes playing a major role in the disease and also against contributory genes found in schizophrenia cases emerging in childhood or the early reproductive years. In this scheme of things, the genes surviving today could be expected to have roles in schizophrenia ranging from slight but significant to very minor—this variation being partly due to the relatively late emergence of some of the more influential genes (leaving scant time for them to be weeded out) and partly due to vagaries of the selection process.

This brings us to nongenetic factors. As we have seen, the role of these other factors may be small. For although a connection has been found between schizophrenia and delivery complications, this connection only becomes really powerful in people with the right genetic heritage. That is, while it looks as though certain delivery complications can trigger long-term development of schizophrenia, this seems to affect mostly those whose genes already make them prone to the disease. This finding does not eliminate all chances that other factors are acting independently, but it does intensify the focus on genetics by raising the possibility that virtually all cases of schizophrenia have genetic roots.

For those who favor wholly nongenetic causes, the big hurdle is delay. That is, if we assume that some nongenetic factor (anything from a virus to the archetypal schizophrenogenic mother) is causing the disease by itself, then we have to explain why the effects are so delayed and why it takes twenty-odd years for schizophrenia to emerge. In most cases we have trouble doing that.

But we have no trouble if we suppose that schizophrenia's primary cause is an array of genetic bugs sometimes assisted by brain-damaging delivery or other factors. For if this state of affairs had previously caused schizophrenia to appear in childhood, the victims would have been unlikely to reproduce, and so most of these particular schizophrenia-provoking genes would have been selected out over the course of human evolution and would no longer be causing schizophrenia. So mostly genes associated with delayed expression of the disease would remain; and whether or not other factors were involved, this genetic scenario would provide a good evolutionary explanation for the delay that we observe.

We know of two mechanisms built into our brains that might help account for the delayed emergence of schizophrenia. One is called "pruning," because just as a gardener cuts back overgrown branches, the brain eliminates masses of neurons at the beginning of the third decade of life. For most of us this maintenance operation is harmless. But for some it might tip a delicate balance in the direction of thought processing problems and schizophrenia.

Also, as we have noted previously, the frontal lobes continue to develop until their owner reaches his or her midtwenties. So certain frontal abnormalities tied to schizophrenia might not get expressed until around that time, when the defective frontal systems come on-line.

Another possible contributing factor is social challenge. In late adolescence we are expected to get a job, leave home, go to college, find a partner, and separate from our families. The long and luxurious dependency of childhood is over, and we have to make our own way in the world. For people barely able to manage the limited demands of home life, this challenge may precipitate disastrous failures—failures that could add marked psychological stress to the lives of people predisposed toward schizophrenia.

Beyond this, we have seen plenty of other long-delay mechanisms built into the genetic arrangements that shape our bodies; and experience with neuromodulators, drug treatments, and mental ills has revealed plenty of powerful delay mechanisms in the brain. So it seems reasonable to think that the expression of many brain-related genes, both sound and defective, is delayed. Of course, with all the existing genetic mixes and variable conditions found in the real world, we should not expect any delay mechanism to be perfect. And in fact we do see occasional cases of childhood schizophrenia, presumably precipitated by failure of some delay mechanism or by some especially perverse combination of defective genes or other factors; but these cases are rare enough to be perfectly consistent with our theory. So all in all, there seems reason to suspect that schizophrenia, despite its many vagaries and variations, may in fact have one primary cause, this being an array of genetic defects lingering like phosphorescence in the wake of the swift, dramatic, and dazzling evolution of the human brain.

If this theory should turn out to be valid, would it further stigmatize schizophrenics as being genetically defective? For several reasons we think the chances are small. First, though the stigma generated by the bizarreness and mystery of schizophrenia is still very strong, it may well be declining now, because we have reduced the mystery somewhat and shown that the symptoms can be treated. If we can explain schizophrenia's basic cause, that may well help to dispel schizophrenia's mystery still further, lessening rather than increasing the social problems that confront its victims.

Furthermore, Adolph Hitler showed us the evils of taking human genetic selection into our own hands, and we have not forgotten. So we tend to be repelled at the thought of rejecting ill people for genetic reasons. Instead, whether the ailment is hemophilia, sickle cell anemia, or something else, we have generally tended to regard the afflicted as people who are sick and need help. So there is no reason to think that fingering a genetic cause would worsen the currently existing stigma surrounding schizophrenia.

Finally, if schizophrenia really is caused by a wide array of genes, publicizing this could have both social and therapeutic benefits. For in that case nearly everyone would likely share some of the responsible genes, and we might be able to reduce schizophrenia's social stigma by making people aware of this shared genetic heritage.

Genetics also opens up a path to prevention. Even now, geneticists can identify families at high risk of having schizophrenic offspring. They find that most structural and other damage arises in the early stage of the illness, around the time that positive symptoms are emerging. And they know that drugs like risperidone, with relatively mild side effects, stand some chance of abating these symptoms. So some theorists, led by Ming Tsuang of the Massachusetts Mental Health Center, are exploring the possibility of preventing schizophrenia by treating likely victims before disease symptoms emerge. In his view, "the scientific tools we have today are so advanced that schizophrenia treatment should be headed strongly into the area of preventive medicine."

This prevention theory is compelling. However, other experts have reservations. They know that even the mildest antipsychotics can have strong side effects; they see no way of selecting only future victims of schizophrenia; and so they suspect that the dangers of pretreating nonvictims would tend to overwhelm potential gains. Also, they have no proof that early psychosis actually causes the brain damage seen in schizophrenia, for it seems just as likely that the damage is wrought by long-delayed expression of hidden genes or other troubles. In that case, early treatment with an antipsychotic might not prevent this damage, and so the benefits of pretreatment could range from marginal to nil.

These reservations are well taken, but it seems to us that the obstacles raised might well be overcome. For one thing, we're getting pretty good at detecting signs of schizophrenia years before any active disease emerges. Videotapes of children who later became schizophrenic have revealed subtle patterns of poor motor coordination in these children, and various other studies have found patterns of poor attention and impaired social behavior. Tests based on these findings have been able to predict psychosis with fair reliability in children of schizophrenic parents. And while these tests clearly need more refining, they can already detect most of those children who will experience schizophrenic psychosis, while screening out most of those who will not.

This by itself should prove helpful for genetic counseling, which should play a part in the treatment of schizophrenia. Wherever possible, such counseling should be provided not only for schizophrenic patients but also for others who are closely involved—including parents, guardians, spouses, and children. For it is frightening not to know who the disease may strike next or what the chances of passing it along may be. And since we know quite a lot about that, we can partially dispel this cloud of mystery.

Meanwhile, our drugs are getting better. So perhaps we are not too far from a time when they might be deployed in a preventive manner. And if they prove unable to keep schizophrenia from doing its initial brain damage, perhaps we can find something else that will. For among other things, we still have a lot to learn about the possible mechanisms that keep schizophrenia hibernating twenty years and more. Perhaps if we learn enough about them, we might find they could be persuaded to hold off schizophrenia indefinitely. That is just futuristic speculation at this point, but it shows that drugs are not necessarily the only answer and that brain science has other prevention-related options to pursue.

But future research is not our focus. Our main concern is proper treatment of current mental ills. In the case of schizophrenia, we know that hundreds of thousands of its victims have been left homeless or imprisoned because they received improper treatment or because whatever treatment they received was poorly followed up. That problem is not scientific so much as social. However, at its core we find a split between drug-heavy biomedicine and humanistic therapy, neither of which can provide effective treatment on its own. Furthermore, it seems clear that the lion's share of recent information about schizophrenia has not come to us through humanistic psychology or pharmaceutical research. Rather, it has come to us through a brain science revolution that has expanded what we know in lots of ways, not just about schizophrenia but also about depression, mania, anxiety disorders, recreational drugs, the workings of sleep and dreams, the nature of human consciousness, and a host of other matters.

The account we have presented here covers only a small part of what is happening in the specific areas named and says nothing about worthwhile advances in many others. But even so, it shows that brain science has advanced enough to provide a foundation capable of uniting the two halves of psychiatry's split personality. Such unification is overdue. We have seen the pendulum of psychiatry's past swing from fledgling biomedicine to ungrounded psychoanalysis to unmitigated overdependence upon drugs, partly because many felt too little was known about the ills being treated to permit anything except blind faith in the fashions of the day. But in fact brain science has learned a lot of what we need to know—plenty for us to improve things a great deal if in fact we can effectively apply this knowledge in a thoroughgoing way to current treatment. How that might best be done is the subject of Part Four.

Prescription for a New Psychiatry

11

Rescue and Resuscitation

The chief difficulty Alice found at first was in managing her flamingo: she succeeded in getting its body tucked away comfortably enough, under her arm, with its legs hanging down, but generally, just as she had got its neck nicely straightened out, and was just going to give the hedgehog a blow with its head, it would twist around and look up at her face with such a puzzled expression that she could not help bursting out laughing.

—**Lewis Carroll,**
Alice's Adventures in Wonderland, **1865**

Psychiatry's current plight is no laughing matter. But as Alice discovered with her flamingo, and as many of us involved in mental health have found with psychiatry, anything really out of whack can be highly vexing. That makes it tempting to talk about psychiatry's demise. But does it make good sense? Psychiatry, after all, is just a branch of medicine whose job is to help the mentally ill; and in contrast to what medicine could do a century ago to help mentally ill people, medicine today can in fact do quite a lot. That means psychiatry's job is important. So if we are to talk about the "death" of psychiatry as something justifiable, we should probably be ready to explain why this important job can be neglected, or else how some other branch of medicine can do it better.

This matter came up in a recent conversation with Daniel Weinberger, a distinguished psychiatrist and neurobiologist at the National Institute of Mental Health. Weinberger affirmed the general view that "Psychiatry is in a woeful state." He also noted:

> Mental illness is a real thing. It's profoundly disabling and profoundly costly emotionally, psychologically, socially, and economically to individuals and society. Psychiatry is the clinical discipline that has taken on the care of mental illness. Whether psychiatry disappears or is taken over by some subpopulation of neurologists or general medical doctors, there are going to be some people who have to care for the mentally ill, and there are certainly unique skills that one needs to learn. . . .
>
> So there are going to have to be clinicians taking care of mentally ill people. Where do they do this? They can do it in HMOs, and they can be like any other doctor in that milieu. But if they are going to practice good psychiatry and be effective in taking care of psychiatrically ill patients they cannot be seeing a patient 15 minutes once every three months.

As this suggests, anyone seeking to resolve psychiatry's current troubles must deal with several problems. First among these is the old dilemma of how to bridge the gap between personal therapy (clinical work with the patient) and biomedicine. Beyond that is how to deal with HMO and insurance company restrictions on psychiatrists, how to structure and coordinate sound care, how to deal with our bizarre laws, and how to recruit people who are ready to bridge the therapy–medicine gap and are also able to confront these other issues.

We shall take up the specific measures needed to deal with each of these problems. But at this point it is worth noting that the problems are quite broad. The problem of how to bridge the gap between clinical work with the patient and biomedicine is something that confronts not just psychiatry but most other branches of medicine. And indeed, the question of how to cope with today's increasingly technical world—how to bridge the gap between our personal lives and modern science—is something that in one way or another affects us all.

Likewise, restrictions imposed by HMOs and insurance companies are doing far more than limiting psychiatry. They are limiting all sorts of medicine. And since they are motivated not by desire for social good so much as by desire for the right sort of bottom line, they sometimes fly in the face of personal needs and social justice. So clearly, this problem bedeviling psychiatry also has far broader implications—as do the problems of how to structure and coordinate care rationally, how to work within the framework of misguided laws, and how to recruit suitably qualified people.

That's not to say that solutions to these problems are easy. But it does suggest that simply caving in and pretending the problems don't exist is not the answer. For sound psychiatry is badly needed. It is needed to oversee sound and affordable care for millions of people with severe mental ills. It is needed to provide effective insight into lesser ills. It is needed to connect psychology and patient therapy with advances in brain science. And it is needed to perform work that no other specialty can do.

This last point deserves particular attention, because today's psychiatry has fallen so far that psychology and neurology seem ready to take over. As noted earlier, psychology is now attracting the "best and brightest" college graduates to its banner, while neurology is enlisting the kinds of medical school graduates that psychiatry attracted forty years ago when Freudian psychoanalysis was king. And high-paid psychiatrists are being locked into pill-pushing dungeons, while lower-paid clinical psychologists have unrestricted access to psychiatry's former therapeutic domains.

That seems fine to many. What makes it seem especially fine is that clinical psychology has considerable appeal to mental patients—because patients are attracted by sound and caring therapy, which is what good clinical psychologists provide. But clinical psychology, directed at therapeutic treatment of the human mind, lacks psychiatry's medical and hard science base. That's critical today, because much of what we know about treating mental ills is medical. The drugs are medical. The knowledge of brain chemistry and brain structures relating to both drugs and the nature of mental ills is medical. But psychology is not medical, and in fact is well known as a refuge for intelligent people who are uncomfortable with hard science. So it is not geared to understanding hard science, supervising medical personnel, or even coordinating with pill-pushers. And despite its promising array of emerging graduates, it cannot reasonably be expected to bridge the gap between our growing scientific knowledge and sound treatment.

Indeed, clinical psychology is not even in a good position to supervise itself. That's because most psychology derives from "top-down" theories based on more or less careful observation of human behavior. These theories have little or none of the "bottom-up" foundation provided by hard science. All too often they are creatures of bright ideas, human whim, or fashion. Like Freud's theories, they have no particular standards to live up to, and they tend to resist proof or disproof. So they all coexist together. That explains why New Age and other gurus have traditionally flocked to psychology; why the field still has a woolly or "touchy-feely" reputation; why all manner of psychology theories are fair game; and why the therapy provided by one sort of psychologist may differ dramatically from that provided by another.

Happily, in recent years this loosey-goosey situation has come under some control. That's because study of the brain has been building a hard science foundation. However marginally, "bottom-up" hard science standards for theories about the human mind have been emerging. So, gradually and indirectly, any psychological theory seeking respectability has needed to show that it does not depart too radically from our growing basic knowledge of the brain. As Daniel Weinberger puts it:

> A lot of hare-brained schemes may be of value. Freud once said that no discipline has a monopoly on the obvious. A lot of the observations made by ego psychologists, psychoanalysts, object relations theorists, etc., were valid observations. The problem is that there has to be a standard by which ideas, theories, concepts are evaluated and tested. That's what I think has been the big change. Psychiatry has been required to operate under a scientific model like other putative health specialties. If there's no standard—if ideas, theories, concepts are not submitted to rigorous scientific examination—then it's just philosophy.

But of course, one cannot get this benefit from psychology alone. It has to be obtained through brain science. And since clinical psychologists are in no position to effectively judge brain science, while properly trained psychiatrists are, that is a good reason to rebuild psychiatry.

Another good reason relates to the nature of neurology. This branch of medicine, which treats brain problems having clearly organic causes (strokes, brain tumors, Parkinson's disease, etc.), has been doing well of late, partly because many bright medical students interested in doing brain science research have flooded in. Moreover, as organic causes for psychiatric diseases like narcolepsy have emerged, neurology has tended to inherit them. As a result, many physicians see neurology as the wave of the future. Indeed, many think it only reasonable, since virtually all mental ills seem to have organic causes, and since psychiatry has fallen on hard times, to make this transition sooner rather than later and simply have neurology take over.

That would be too bad, for neurology lacks the therapeutic side. Like most doctors, neurologists must respond to patients' needs and have a reasonable bedside manner. But they have little or no special training in psychotherapy, nor are they prepared to deal with chronically irrational patients, or with patients needing prolonged therapy, guidance, counseling, monitoring, social services, and follow-up. What's more, though neurology appeals to those who like the scientific side of medicine, it looks forbidding to humanists whose forte is being unusually sensitive to human needs. In other words, neurology and psychology each lack what the other has; while psychiatry, if it were working properly, would serve as an effective bridge between them.

But the strongest argument for rebuilding psychiatry relates to treatment. Even treatment of minor mental ailments may be poor these days, because the general practitioners, therapists, and pill-pushing psychiatrists involved may be inept at diagnosing psychiatric ills and devising sound treatment. But this problem is grossly compounded, as we have seen, when the patient is severely ill, occasionally or usually psychotic, and in danger of being lost to treatment.

This means we need a mental health system led by psychiatrists who have a good command of biomedicine and therapy; one that includes psychologists, therapists, counselors, social workers, and others assigned tasks corresponding to their skills; one that can hospitalize patients long enough to devise effective treatments and see they are taking hold; one that can provide clubhouses and other sorts of supervised living arrangements; and one that will keep tabs on patients to ensure they are not lost to treatment.

One of us (Jonathan Leonard) discussed this matter with Dr. Steven Hyman, Director of the National Institute of Mental Health, in August 1999. His view:

> I have no crystal ball, but I think that for many, many situations we have long passed the point at which the situation becomes unacceptable for many patients. We are now in many cases doing a disservice to health care. . . .
>
> Especially for a person who has some impairment, these fragmented, underfinanced systems of ours often really are so difficult. I mean, if someone with schizophrenia doesn't show up at their appointment, in some settings they will get a postcard saying "You're terminated from the clinic." But these things have been studied; there's very good data; and there are assertive community treatments. Indeed, there are different forms of assertive community treatment; and when someone doesn't show up, then it's time to go out there and find them and find out what's going on.

So there seems no doubt that psychiatry should be rebuilt. The question is whether it can be. For as Judith Rapoport and others have pointed out, many of those now entering psychiatry seem marginally qualified to prescribe drugs and even less qualified to care for patients. And many of the psychoanalytically oriented psychiatrists who preceded them from midcentury onward chose almost purely therapeutic careers divorced from medicine. Given these deficiencies, it seems reasonable to ask whether the field has been hopelessly depleted of qualified bridge-builders.

At least for now, the answer to that question is clearly "No." To begin with, not all of the new psychiatric residents are hopeless. Some people who finish toward the bottom of their medical school classes turn out to be good doctors. Some bright foreign graduates, even from places as far afield as China, succeed in overcoming cultural and language problems that interfere with their careers. And some experts now feel that we have weathered the worst of the recruitment crisis and that, for whatever reasons, the general quality of new residents is improving so that today such top-flight programs as those affiliated with Harvard, Columbia, Brown, University of California–San Francisco, New York University, Cornell, and others still have no trouble filling their residency slots with bright, energetic, and highly motivated young people, particularly those interested in public sector psychiatry or in neurobiologic research. Not coincidentally, it is in these areas of community psychiatry and research that American society is least ambivalent in its attitude toward psychiatry.

Beyond that, many of the psychiatrists belonging to the throng once steeped in psychoanalysis are still around. Many of them didn't come from the bottom of their med school classes but from the top. And while some wholeheartedly embraced psychoanalysis and clung to it as the world turned, others were more flexible. Indeed, the full range of adaptation found within this group is startling.

Some of those trained to follow Freudian precepts grew suspicious or rebellious. One of us (Allan Hobson) interrupted a Mass Mental residency for a brain research job at NIMH, believing that psychiatry had no business being divorced from medicine. Daniel Weinberger, for similar reasons, completed his Mass Mental residency but then did a separate residency in neurology. Joseph Schildkraut, another Mass Mental resident, stayed on as a Mass Mental staff member and became noted for his sensitivity to patients; but he also did landmark medical research on depression.

These are just a few examples of a far more general trend. One of a student's best tools is healthy skepticism. And since psychoanalysis had real trouble delivering on its claims, while drugs and brain science were moving forward, large numbers of medically trained psychiatric residents, researchers, teachers, and practitioners became involved with biomedicine and also with nonanalytic forms of therapy. Even the analysts—who accounted for only 10 percent or so of all psychiatrists, who had been trained in the psychoanalytic institutes, and who had made a clear professional commitment to analysis—began changing their commitment as the analytic boat sank. As we have seen, by 1974 three-fifths of all analysts polled by the American Psychoanalytic Association had begun prescribing medications and a third were offering nonanalytic therapy.

Since then this trend has accelerated. One bridge-builder with a strong background in this area is Mark Solms, a young British psychoanalyst who has a private analytic practice and also belongs to the Department of Neurosurgery at a teaching hospital in London. As he explained in a recent conversation:

> Over the last thirty or forty years there's been enormous progress in the neurosciences. Psychoanalysis has remained insulated from that progress. So there's a catch-up job to be done by psychoanalysis, and I'm interested in participating.
>
> In the past there were obviously many reasons why virtually all psychologies, among them psychoanalysis, were prepared to have such slender bases for their theories and didn't feel the need to test them against competing theories. But one of the main reasons was a lack of appropriate methodologies for doing this. And so there was, to put it unkindly, an excuse for proliferation of all sorts of theories and all sorts of treatments which had more or less questionable scientific bases.
>
> But that excuse no longer applies. The advances in the neurosciences, methodologically speaking, have been so enormous, and that's also indicated by what's happening in neuroscience today, that neuroscientists are interested in the very same functions that traditionally were the preserve of the psychologists and the psychologies. Things like dreaming and consciousness and emotion and so on. So there's absolutely no reason why untested psychological theories should remain insulated from scientific progress any longer.

No armchair theorist, Solms is coeditor of an ambitious new journal called *Neuro-Psychoanalysis* that seeks to have neuroscientists and psychoanalysts communicate with one another and work together on problems of common interest. Despite his London base, he has also become the director of a center established by the New York Psychoanalytic Institute for interdisciplinary psychoanalytic and neuroscientific research. The center represents an unusual breakthrough because, as Solms notes:

The New York Psychoanalytic Institute is the oldest psychoanalytic institute in the English-speaking world and also might be regarded as the most isolationist. Traditionally they're seen very much as the most conservative Freudian institute around. But over the past decade they've had an amazing change of heart, born no doubt of the crisis faced by psychoanalysis and recognition that if psychoanalysis is going to survive it has to do something radically different. So they've taken the bull by the horns and established this center, appointed well-known neuroscientists to its advisory board, and begun activities that include small-group research projects and monthly presentations by prominent neuroscientists.

How successful this endeavor will be is hard to tell. But success is not the point. The point is that the decades-long trend away from "pure" psychoanalysis and toward brain science has reached into even the most conservative recesses of the Freudian citadel. So clearly, we are not looking at one group of psychiatrists who are therapists and another group who are pharmacologists. Instead, we are looking at a broad spectrum of people, many of whom have strong backgrounds in both pharmacology and therapy, all of whom were comfortable enough with natural science to get an M.D. at some point, and many of whom have made professional adjustments that involve far more than a nodding acquaintance with advances in brain science.

What makes this most encouraging is that in our call to rebuild psychiatry we are not searching for a great mass of psychiatrists to mount the reform barricades. We have no wish to see everyone with any degree of mental ill analyzed for up to five hours a week. Our desire is much more modest. We want to see most patients treated rationally. So we hope to see biomedically and therapeutically qualified psychiatrists take enough time with mental patients to understand the patient's problem, establish a good therapeutic relationship, devise a sound course of treatment (drugs, therapy, and other measures), arrange for follow-up, and arrange to review the case or see the patient again after a specified interval, typically a week to a month from the time of the last visit.

Clearly, this is not happening in most places. So if it suddenly started happening for, say, two million neglected people, there would soon be a shortage of qualified psychiatrists. But transformations like this take time. We didn't empty the mental hospitals in a day, and we cannot undo the current foul-up in a day. So if we start developing comprehensive treatment programs that make sense, we will not run short of qualified people. Instead we will find a fair-sized reservoir of qualified psychiatrists on which to draw.

Beyond that, such treatment programs would begin raising the interest of medical school graduates in pursuing careers in psychiatry, and would begin generating a qualified body of new talent. For if psychiatry is properly practiced today, it offers three compelling benefits: a tie to our growing knowledge of the brain, an opportunity to relate personally to patients, and a chance to do sick people a lot of good. What turns off medical school graduates today is the fact that new psychiatrists are forced into a drug-prescribing mode and have no chance to properly practice their profession. If this were turned around, and if psychiatry were allowed to be productive, recruitment would turn around as well.

Of course, all this will cost money. But as our HMOs, insurance companies, and hospitals are finding out, improper care can prove more expensive than sound treatment. For instance, a lot of labor goes into hospital admissions because doctors, therapists, and nurses need to know something about anyone they treat; and of course, any stay in a hospital is expensive. So a severely ill mental patient sent through a hospital's revolving door and discharged after a few days racks up heavy costs. But this treatment isn't likely to help much. So in short order the revolving door of this or some other hospital turns again to admit and discharge the same patient. In many cases these short and futile journeys wind up costing more than a single course of sound treatment that can stabilize the affliction and improve the patient's lot.

In this same vein, a marginally qualified psychiatrist who sees a patient for fifteen minutes can't do much; nor can a nonmedical therapist coordinating only marginally with the psychiatrist; nor can mental health workers lacking effective central coordination. So while all these various wheels may turn and incur their respective costs, they are less likely to help the patient than soundly coordinated services whose agents are given enough latitude to do their jobs well. True, doing the job well may mean higher initial costs, but the affliction is less likely to continue in high gear and the patient is less likely to keep bouncing around from one high-cost caregiver to another. So doing the job well may wind up being cheaper than what we all too often are doing now.

People learn from experience, so HMO administrators may come to recognize the importance of sound psychiatry. Or they may not, especially if they are too intent on their mission of cutting costs by shaving services. In either case, however, the HMO problem may prove less of a long-term obstacle than many think, partly because even HMO administrators are adaptable and partly because the HMOs are in deep financial trouble.

The reasons for this trouble seem clear enough. Medical costs are very high. The HMO adds another layer of bureaucracy atop the health establishment to cut corners. And HMO employees get paid plenty, because you can't hire competent people to do a nasty job like depriving others of health care benefits unless you pay them well.

Even so, the HMOs have had limited success cutting corners, largely because of doctors. Unlike doctors of yore, the earning prospects of today's doctors look pretty dim. So most people don't become doctors today to make money. They become doctors because they want to help the sick. The HMOs interfere with that. So many doctors have no qualms about misrepresenting cases—reporting the wrong ailment, completing the wrong form, and thinking up thousands of evasions—in order to procure sound care for their patients.

It's not hard to imagine how this monster/superhero struggle is likely to come out. The HMOs may yell, scream, and sue the doctors. But they won't get much sympathy, because ours is not an impoverished country. Ours is the richest country in the world. So most people—including most judges and politicians—will remain quietly sympathetic with the doctors; many HMOs will continue to go broke; and in time just about everyone will recognize that the whole HMO concept is fraught with financial peril.

This suggests that a sea change could be coming, and that today's health system could differ greatly from whatever we find in place a decade hence.

But even if it does, that will not necessarily solve psychiatry's problems. For those problems do not arise mostly from evil HMOs or evil laws or evil anything. At heart, they arise from a misunderstanding—from a belief that psychiatric therapy is a paper tiger and can be effectively replaced with biomedicine.

In a way, brain science has fueled this misunderstanding. For as the Freudian smoke cleared, anyone could see that psychiatric drugs worked and also that brain science was making progress. So, for the general public, it seemed reasonable to think that one and one made two, and that the true answer for mental ills was psychiatric drugs.

As we have seen, however, the real message for professionals involved with brain science was very different. The drugs help, all right, but they are limited. They don't always work, they don't cure anything, they often reduce but don't eliminate the target symptoms, they affect the whole brain's chemistry, they can have nasty side effects, and they definitely won't work if the patient doesn't take them. For all these reasons, it's important that drugs be prescribed by someone with a good understanding of the patient's case and circumstances. And of course for patients with severe mental ills that's only the beginning—because it is necessary to ensure that the drugs are taking hold, to establish a sound therapeutic relationship, to provide effective monitoring, and to coordinate actively with appropriate social services. So, while drugs commonly play a key role in all of this, they are by no means any cure-all.

If that's so, why are they being treated like a cure-all? The real answer, when all the layers of the onion are stripped away, is that once people lost faith in the Freudian rationale, they stopped believing that humanistic therapy was important. That's because the Freudian rationale provided a sacrosanct foundation for both psychoanalysis and the therapy practiced by most psychiatrists. So when that rationale was exposed as guesswork and swept away by brain science, there was no obvious way to replace it; and then the public's growing fascination with tranquilizers and other psychiatric drugs swung the pendulum over to biomedicine.

As the history of psychiatry shows, this swinging of the pendulum didn't happen overnight. It took decades. But even so, it may have gone too fast, because it looks like we missed something important. We missed the fact that brain science provides no ringing endorsement of biomedicine's victory over psychotherapy and in fact is highly compatible with them both. Sigmund Freud sought such a connection between science and therapy over a century ago, when he tried to devise a psychology based on the brain science of his day. When that failed, he resorted to guesswork like everybody else and came up with a compelling if faulty rationale for one sort of psychotherapy. Now the scientists in their white coats have caught up with Freud and have produced knowledge invalidating much of his guesswork. But that doesn't mean that Freud's rationale should be replaced with nothing, or that therapy can be replaced with drugs. Rather, it means we now have much of the knowledge that Freud originally sought but couldn't find, and so we can begin to construct what he originally wanted—a bottom-up psychology to guide therapy that is based upon brain science.

12

Neurodynamics: Toward a New Psychology

Our first conclusion. . . is that a certain amount of brain physiology must be presupposed or included in Psychology.

William James,
The Principles of Psychology, **1890**

We have quite specific ideas for building a new psychology, and we will get to them momentarily. But it would be the height of folly to work in a vacuum—to proceed without considering what psychiatrists and most psychologists now use as their principal psychology. That psychology, known commonly as "psychodynamic" psychology, derives mostly from Freudian psychoanalysis. So before defining the principles of our new neurodynamic psychology, we shall take a closer look at the power of psychoanalysis and the pros and cons of its psychodynamic offspring.

While people have been pretty hard on the Freudians in recent years, we should recall that the smugly self-confident Freudians did some things rather well. They pointed out the importance of unconscious processes; they highlighted the role of emotions in mental ills; and they employed a form of therapy that roused widespread enthusiasm in many of its practitioners and kept multitudes of mental patients coming back for more.

That's probably why an abridged version of Freudianism still flour-
ishes. As we have seen, psychoanalysis—the costly, intense therapy in
which the patient makes very frequent (often daily) visits to a psychoanalyst and lies on a couch from where he cannot observe the psychoanalyst—
is in limited demand. But less intense forms of "psychodynamic" therapy
derived from psychoanalysis are widely practiced. In fact, psychoanalysts
have long trained innumerable psychology students and psychiatry residents in psychodynamic therapy, and they still do—which suggests that at
least some elements of that therapy must be worthwhile.

One important element of psychodynamics, perhaps the most important,
is the principle that the therapist must empathize with the patient. That
means not being judgmental. It also means trying to feel what the patient
feels, getting on the same emotional wavelength, and seeing things from the
patient's viewpoint.

The therapist's goal in using this approach is to build a relationship
where the patient feels sufficiently safe and trusting to learn from the therapist. To do that, the therapist must learn how to act almost reflexively in the
patient's interest. As noted anthropologist Tanya Luhrmann points out in
Of Two Minds: The Growing Disorder in American Psychiatry, a book that
explores the training of psychiatrists, the therapist "must be able to respond to a patient according to a patient's needs rather than his own. He
must be able to listen to a patient without being caught up in his own embarrassment, fear, desire. As one supervisor remarked to me, 'When the patient says "You are a fascist," the therapist must be able to say, "How am I a
fascist?" To explain to him that she [the therapist] is not a fascist serves her
own needs. To understand how she appears to be one serves his.' "[1]

This approach looks a bit like double-entry bookkeeping. But bizarre as
the process seems, it actually bears a strong resemblance to something
many people do every day, which is parenting. For ideally, parents do empathize with their children. They try to understand. They are emotionally
attuned. And they do act as authority figures, provide a safe haven, seek to
have their guidance accepted, and typically put their children's needs before their own. Thus, so long as the parenting is good, one can hardly imagine a parent whose child says "You're a fascist" replying with anything but

"How so?" or "Why do you say that?" or some other answer equivalent to the psychodynamic therapist's response.

The approach turns out to be useful, because a parent-type relationship is precisely what multitudes of the mentally ill really need. They don't entirely trust their own minds, so they have trouble trusting others. They may have no one to talk to about their problems, especially problems with strong emotional overtones, and they need expert guidance. Moreover, besides needing someone to trust, they need someone who they believe understands their problems. So ordinary friendship won't do. Nor will an impersonal professional relationship, even if it yields good advice, because the advice will not sink in. What the patient needs is an expert empathizing like a good parent, and that is what today's psychodynamic therapy typically provides.

Of course, therapists who follow Freud's injunctions against emotional involvement cannot really provide parenting. There is an intrinsic conflict between the psychoanalytic ideals of neutrality, objectivity, and distance and the child's need for commitment, belief, and closeness. In fact, Freud was paternalistic and condescending rather than truly parental. And analysts who followed his lead in this regard were more committed to analysis of parental transference than to parenting as such. Even so, most analysts became de facto parents whether they liked it or not. That is because the daily sessions, the couch, the informal clothes, the homelike setting, the focus on repressed childhood memories, the emphasis on family relationships, and the promotion of transference all served to encourage and reinforce a bond between the psychoanalyst and patient that was essentially parental.

The result of all this, as full-blown psychoanalysis emerged, was quite compelling. Indeed, the Freudians had found a magic spell that worked on just about anyone who was receptive. Besides working on the mentally ill, it worked on others who were not ill. That's one reason the Freudians ascribed a "little bit of mental illness" to just about everyone and felt everyone could benefit from analysis. It also explains why future analysts in training are still expected to undergo analysis, and why they generally find the experience overwhelming. As Tanya Luhrmann points out:

Therapy relationships [in psychoanalysis] are emotionally intense in ways that are quite incomprehensible to an outsider. A resident [in psychiatry providing psychoanalytic therapy] has some patients who love him, others who loathe him, and some who threaten to kill themselves when he goes on vacation. Many of his patients cry copiously into his Kleenex. Sometimes he buys Kleenex by the case. When he is in psychotherapy, he too weeps copiously, apologizes for it, and then weeps some more. Young therapists are often taken aback by the strength of their own and their patients' feelings. Some of them make decisions about where to live on the basis of where their analysts live. "My analyst is unwilling to relocate to San Francisco [to which this resident had planned to move at the end of residency]. Well, I like this city, even if it isn't San Francisco. So for now I'll stay." Or, as one resident more simply said about his analyst, "God, I like him."[2]

Nobody has ever figured out just how this works. What we do know is that Nature has made the emotional foundation of the parent–child bond very strong, and it seems clear that Freud's followers found a way of tapping into that. But just because full-bore psychoanalysis is overwhelming doesn't make it an ideal treatment for mental patients. On the contrary, to all appearances its intensity works against it—causing it to set up a highly artificial relationship that encourages dependence, raises emotions, and in some cases proves counterproductive. Indeed, nearly a century of psychoanalysis has failed to demonstrate decisive benefits, and by today's standards its cost, intensity, and duration look excessive. What's more, psychiatric drugs and brain science have provided other tools and an understanding of the mind that are very different from anything Freud imagined. So it seems reasonable, as the Freudian psychology tied to repressed childhood experiences and family relationships continues to retreat, that psychoanalysis should be limited to those few who still want it and can afford to pay.

But less intense psychodynamic therapy is clearly helpful. Reasonably spaced sessions, typically starting at one per week and tapering down to one per month or so, can improve a patient's sense of security, provide a sounding board for the patient's thoughts, open up a conduit for expert guidance, and grant the therapist effective access to the patient's mind. Beyond that, psychodynamic therapy tends to keep mental patients coming back to the therapist. That's important, because many mental patients today are lost to treatment, and a key element needed for effectively treating mental ills and coordinating the various services involved is continuity.

Sadly, the retreat of Freudian psychology and arrival of the HMOs has left psychodynamic therapy stranded like a beached whale. No longer can therapists claim to cure most mental ills by discovering repressed childhood memories or by working through problems arising from schizophrenogenic mothers. They can now wave their arms and talk vaguely about patients who need reparenting because they were abused or neglected as children; but this is neither very credible nor very convincing.

Unfortunately, biomedicine has made no serious moves to replace the lost Freudian mystique. Rather, it has tended to treat mental ills like other physical ailments. While this has provided some justification for what biomedicine is doing, it has left biomedicine stranded and isolated much like psychotherapy. In other words, failure to bridge the gap between the brain and the mind has meant that the now largely discredited psychology devised by Freud and his followers has never been effectively replaced. So that psychology, which swept in like some great tide, has simply swept out again, depriving everyone from doctors to politicians of a convincing rationale for sound treatment of mental ills and condemning psychiatry, therapy, and vast multitudes of the mentally ill to a prolonged era of vulnerability and neglect.

The time to end that era has now arrived. But to do it we must be properly equipped. And since public confidence in Freudian psychology has ebbed, one critical piece of equipment that we need is a new psychology for psychiatry, a psychology linking the brain and mind and showing how biomedicine and psychotherapy should work together. Naturally, one could dedicate an entire book or even several books to this purpose. But having laid out much of the necessary groundwork in the preceding chapters, we feel reasonably well prepared to suggest what the guiding principles of such a psychology should be.

The fundamental point is that this psychology should be compatible with what cognitive neuroscience and other branches of brain science have discovered about the mind, and also agile enough to adapt to new discoveries. It should take advantage of experimental opportunities for examining the mind's work—in fields as diverse as experimental psychology and magnetic resonance imaging—and it should warmly embrace a truly scientific approach to verifying observations.

This new psychology should also embrace certain ideas coming in from psychoanalysis that are compatible with what we have learned about the mind. These include the status that psychoanalysis accords to unconscious mental processes, its focus on emotions and instinctive behavior, and its emphasis on the felt experiences of the individual. Within this context, the new psychology should deal with many things that traditional psychoanalysis deals with, like dreams and delusions, but should exclude theoretical explanations of these things that conflict with current brain science or that defy scientific evaluation.

In other words, as noted earlier, we disagree with treating the mind or "psyche" like an independent entity not subject to scientific assessment, even as we agree with the stress psychoanalysis has placed upon dynamic mental processes.

What fresh ideas can our new neurodynamic psychology bring forth? That depends largely on the particular interests of those heeding our call to arms. We have not the space, inclination, or desire to cover here anything like the whole mind science territory that neurodynamics opens up. However, we can single out certain ideas that we regard as fundamental, and we can sort these into several groups—one group being concerned with the healthy mind, another with the nature of mental ills, and a third with the treatment of mental ills.

We shall discuss the healthy mind first because the mind's normal activities help set the stage for dealing with mental illness; and we shall take up mental illness next for a similar reason, because that helps pave the way for discussing treatment. Many of the ideas presented have been touched on earlier in the chapters on brain science. We restate them here because it's important to explain how they relate to the mind as well as the brain, and how we can use them as part of this new psychology dedicated to treating the mind and brain together.

Certain concepts that deal mostly with the healthy mind are as follows:

The mind arises from the brain. This is a very old idea. Indeed, both Sigmund Freud and William James, the father of American psychology, recognized that the brain and mind constituted an integrated system.

Of course, today things are a lot clearer. We now know where the processing of everything from visual inputs to higher-order thought occurs within the brain. We know how the brain generates a wide range of feelings—elation, satisfaction, anxiety, fear, sadness, and others—as well as mental states that include waking, deep sleep, REM sleep, dreaming, and various psychotic states associated with drug abuse or mental ills. We can see that all mental acts from serving tennis balls to playing chess arise from corresponding brain actions. We know the brain has a vast capacity appropriate for performing the mind's full range of activity. We understand that damaging the brain damages the mind. We perceive that without the brain there isn't any mind. And of course, we have never detected anything else that creates the mind. We can also appreciate the philosophers' old point that an immaterial nonphysical spirit would have to do material physical work—and thus become detectable—if it wanted to generate the mind. And we can see that such intervention appears unnecessary, because the brain shows no signs of needing help. Therefore, it is reasonable to conclude that the brain generates the mind.

The mind is divided into functional compartments. As noted earlier, the brain is divided into myriad small specialty shops dealing with vision, hearing, language, speech, smell, taste, touch, movement, various emotions, several kinds of memory, consciousness, attention, thought, judgment, and so forth. Therefore, a mind engaged in reading will use at least some different parts of the brain than a mind engaged in swimming, and both will use different brain areas than a mind engaged in escaping danger or making love. For this reason the mind gives every outward appearance of being functionally compartmentalized, and in fact it is.

Mental acts require coordination. It stands to reason that since the brain is divided into many small specialized sectors, practically any mental act worth doing requires coordination. For example, the decoding of incoming visual signals requires coordinated processing by many of the brain's vision-related specialty shops. Registering the results of this processing in consciousness requires coordination with the thalamus. And getting a refined sense of what the results mean entails coordination with many other regions. Thus, almost any mental activity involves lots of coordination.

The mind's state depends on the brain's chemistry. We have seen how natural alterations in brain chemistry play a large role in the sleep–wake cycle, REM sleep and dreams, the fight-or-flight response, sadness and grief, and a variety of other mental states. So without belaboring the obvious, one can conclude that brain chemistry strongly influences both brain coordination and the mind's state.

The conscious mind is limited. Consciousness, as we have previously defined it, is awareness of information processed by the brain. When your conscious mind performs some act like reading this sentence, it becomes aware of what the sentence means. It isn't aware of all the operations performed on shapes, letters, words, grammar, syntax, and context to extract that meaning. What's more, while your conscious mind was dealing with the sentence it was ignoring a host of memories, procedures, feelings, and incoming sensory signals to which it could have been attending. So even though the conscious mind's attention can shift about, it seems obvious that the share of overall brain activities registering in consciousness at any given time is rather small.

The conscious mind acts as a unit. We perceive that our conscious mind acts as a single unified entity, despite the fact that it arises from a multitude of individual neurons whose identity shifts and changes within the brain from one moment to the next. Our subjective belief that this unity actually exists is supported by an array of computer, split-brain, and other studies referred to earlier which tend to affirm that recurrent parallel firing of neurons (such as that occurring in the thalamus) can bind the activities of different specialized neuron groups into a coherent whole, making them operate together temporarily as a unit.

Most brain activity is unconscious. Redefining the unconscious to acknowledge this fact brings about an important shift in emphasis. The unconscious mind is normalized rather than pathologized. It appears as an asset to be appreciated rather than a liability to be feared. And even its most peculiar manifestations, like dreams and fantasies, have a normal, health-promoting function rather than a debilitating symptomatic one. In taking this position we subscribe to John Kihlstrom's concept of "the cognitive unconscious" and celebrate the bountiful talents that our brains apply to activities outside of our awareness.

The conscious and unconscious minds are friends. If the unified conscious mind is limited, as we know it to be, and if we consider all brain activities to be part of the mind's work, that leaves a very large scope for unconscious activities. But the unconscious mind is not some sort of repressed personality or alter ego like that envisaged by nineteenth-century theorists. Rather, it consists mostly of basic mental processes that the conscious mind doesn't need to know about in order to manage what is happening. Of course, the unconscious mind includes a lot of potentially influential elements—including survival-directed messages hard-wired into the amygdala; biological drive mechanisms that cause the conscious mind to feel a variety of needs; and various clocks, signalers, and switches that can shift the brain's neuromodulator balance. Even so, the brain is not arranged in any way that seems calculated to create an alter ego. This implies that the unconscious mind is not set up to be some sort of antagonist, but instead is a powerful friend of conscious processes, bound by a shared interest in survival and dedicated to providing the conscious mind with whatever the latter needs to do its work.

As noted earlier, neurodynamics is not limited to basic concepts and in fact can work at any level desired by its users. So before proceeding on to the nature of mental ills, we would like to make a slightly more involved use of this approach in the specialized area of sleep and dreams.

For starters, we have seen that dreams are not coherent unconscious wishes cut by a hidden censor as Freud imagined. Rather, the brain assembles dreams like a crazy quilt from mismatched bits and pieces of information. The mind may dredge up old memories in dreams, because our dream-laden REM sleep seems dedicated to learning consolidation or problem-solving that involves binding old and new memories together. But we don't need any professional interpreter or decoder to tell us what our dreams mean, because our brain has already done its best to produce transparent meaning consistent with our felt emotions at the time. So if we look at portions of a dream that we recall, we can often see that these recollections provide insight into our emotional state and other issues of concern at the time we had the dream.

I (Allan Hobson) remember a very recent dream in which I was being actively seduced by an irresistibly attractive younger woman. She was so beautiful that I could hardly believe her ardor in pursuing me. This caused me to be reserved in my response, as if to test her motivation and determination, but not so immune to her charms as to be definitely unavailable. I can still see (and feel) the warmth of her dark hair and eyes, the intense redness of her lips, the clarity of her gaze—all of which invaded my being and flooded me with desire.

What had begun as a private dialogue seamlessly switched to a public setting where I was a member of an audience to which my female pursuer was addressing an impassioned speech. She explained that she was raising funds for a cause so compelling that she didn't even need to describe it. She then added that she had offered to make love to anyone who would donate $300, and that to her great pleasure I had pledged that amount!

Despite the fact that I had no recollection of any such negotiation, I felt pleased to have resolved the issue of motivation to our mutual satisfaction. And the next thing I knew we were alone again and eager to proceed.

But now a disquieting concern entered my mind. How in the world could I consummate this delicious romance while wearing the Foley catheter that had been inserted through my penis to relieve the urinary obstruction caused by my enlarged prostate gland? I had no choice but to inform my would-be lover of this obstacle. To my great relief she seemed tolerant, even supportive, of my indisposition, which we both felt sure was only temporary.

The meaning of this dream seems transparently obvious to me. Because I am now facing imminent surgery to remove my hypertrophic prostate gland, I am more than usually concerned about my sexual attractiveness and competence. But before discussing this "interpretation" of the dream's content, we need to emphasize how much of the dream can be explained by a neurodynamic analysis of its features, namely:

Hallucinoid imagery. The picture of my seductress was so clear and vivid that I can still see her. No doubt about it, she was shot in technicolor. And she and I continually moved through the dream space in our dance of mutual attraction. These formal dream features are ascribed to activation of the brain's sensorimotor, emotion, and instinctual centers.

Delusional acceptance. From start to finish, I had no doubt that I was awake. The highly unlikely sequence of events did nothing to alert me to the fact that I was dreaming. Self-reflective awareness and critical judgment were gone, except for the unusual recognition in scene three that I was in fact sexually unavailable because of the catheter. We ascribe these dream features to deactivation of the frontal cortex and resulting impairment of executive functions.

Bizarreness. There was less than the usual discontinuity and incongruity in this dream. But the scene shifts, from an intimate to a public setting and back again, are typical. Also typical is the ad hoc explanation of my dream woman's sexual motive, a worthy cause so obvious that it was unspecified! In addition to frontal cortex deactivation, we suppose that a combination of serotonin/norepinephrine demodulation and acetylcholine-based hyper-stimulation contributed to the dream's jumbled plot.

Emotion and instinct. The mix of sexual attraction and arousal related to both the hypothalamus (whenever I am abstinent I have more sexual dreams) and the cortex (I have been consciously frustrated by my disability). And the unconscious fear of a further loss of sexual capacity contributed as well.

Memory. My recall was good because I woke up promptly and noticed the details. It was early morning, when my brain was reapproaching waking, so my arousal was sharp. I recounted the dream to my wife (to whom it was obviously dedicated anyway), and that helped me etch it in my memory. Then I got up to empty the urine bag attached to my catheter.

What else is there to say? This dream is all about the hopes and fears of a 67-year-old man with a 43-year-old wife who is enduring the forced sexual abstinence of catheterization and facing the threat of more permanent disability due to imminent prostate surgery. Yes, there are wishes. He hopes he is still attractive. And he hopes he will be accepted and supported through his upcoming ordeal. But there is no disguise and no censorship that we can see. Why the dream woman is not specifically his wife may seem mysterious. But such personal identity confusions are typical of dreams, and it seems to us that they reflect organic dream-related difficulties with character and face representation rather than psychologically motivated "displacements."

By relating this dream, we hope to show how our theory works at the level of understanding dream processes, and how each dream is more an open and immediate appeal for acceptance and understanding than a cryptogram needing decoding to reveal remote and abstract motives.

More generally, we can also see that dreams tell us a lot about the mind. They give us a window on it; they let us see it in a natural state very different from its normal waking state. And what we see is fascinating. For we see a mind in no position to cope with reality. Its thought, judgment, and ability to remember have been dimmed. It accepts all sorts of illogical, bizarre, mixed-up things as being real; fails to exercise critical judgment; and then forgets most of what it saw. But when we collect dream reports by interrupting REM sleep, we find that the reported dreams are commonly laden with strong feelings, usually fear or anxiety, indicating that emotion centers, most notably the amygdala, are typically very active, and that these feelings are woven into dream plots in a way that faithfully reflects actual concerns.

As previously noted, the mind is conscious in REM sleep. So here is a natural state of consciousness in which the mind is seeing things, is emotionally overwrought, and is unable to deal with reality. Brain imaging studies of dreamers in REM sleep affirm this assessment. They show reduced blood flow to an anterior (front) part of the brain called the dorsolateral prefrontal cortex, which is believed responsible for working memory and directed thought; and they also show increased blood flow to the amygdala and other emotion centers below the cortex.

Interestingly, imaging studies have found a similar blood flow pattern during waking when strong negative emotions are being felt. So there seems to be a reciprocal relationship. When the amygdala is highly active the dorsolateral prefrontal cortex is suppressed and vice versa. The result: In dreaming, as in an anxiety attack, it is impossible to think rationally unless deliberate training has prepared one to reassert cortical control over the subcortical emotional process. In the case of dreaming this shift is called lucidity. In the case of anxiety it is called getting a grip on yourself!

This crude but neurologically specific model has obvious relevance for building a unified theory of dreaming and emotion. What's more, it accounts for many of the things psychoanalysis tried to explain with much more complex and cumbersome psychic mechanisms. The following are among the more important changes in psychodynamic theory required by this neurodynamic model:

- Negative emotions like anxiety are not always symptoms of conflict.
- Anxiety arises from natural brain activation and has the easily understandable function of promoting caution, wariness, flight, and related behavior. So within limits, anxiety is normal and clearly useful.
- The reason why dreaming is commonly characterized by anxiety is because the amygdala is activated in REM sleep.
- The reason why dreaming exhibits discontinuities, incongruities, limited self-awareness, and failure to direct thought is because the dorsolateral prefrontal cortex is deactivated in REM sleep.
- Functions of dreaming suggested by these findings include maintaining the capacity to generate adaptive anxiety, consolidating recent learning, and reorganizing memory.

More generally, dreaming is only one of a group of dissociative mental states (including such things as hysterical paralysis, loss of attentional focus, and hypnosis) that have long been scrutinized by psychodynamic and other theorists. In the future, there seems no reason why these matters should not be explored more fruitfully through neurodynamics.

As previously noted, however, our immediate goal at this point is more basic. Therefore, we shall now turn our attention to a number of fundamental neurodynamic concepts relating to the nature of mental ills. These are as follows:

We need to be more critical about childhood memories. We know that explicit memories (memories of facts and events rather than memories of how to do things) get stored first in the hippocampus. Then, as time passes, they are shifted out of the hippocampus and scattered about the cerebral cortex like pieces of a mighty jigsaw puzzle.

These memories can be recalled in various ways. Some can be recalled immediately upon demand. But the brain also has a search mechanism for cajoling reluctant puzzle pieces out of hiding. For instance, if you cannot remember someone's name at a moment's notice, you can activate the brain's search mechanism—not by forcing the issue but simply by registering interest in that forgotten name—and in due course (anywhere from a few seconds to the next day) you stand a fair chance of having that name pop into your conscious mind unbidden.

Of course, if you are able to recall things closely associated with your forgotten memory that can help the search process, because it gives the brain a better idea of where to look. Last week, as of this writing, I (Jonathan Leonard) happened to remember a child's book suitable for my two-year-old daughter Maria entitled *Each Peach Pear Plum*. This book's very brief text is actually a 26-line poem. I couldn't recall many words to that poem at first, having last seen the book ten years ago; but recalling some words helped me to recall others, and within a day or so I could remember the whole thing—accurately as it turned out when I checked this recollection against a new copy of the book.

But poetry is self-correcting to a degree because it rhymes. Not all memories are poetic, and as the memories to be recalled get more complex and the edges of the scattered and faded puzzle pieces become tattered, accuracy becomes a major issue. For instance, my father had a distant relative who was old when he himself was young, and whom he visited several times as a child. She had been born some years before the Civil War and was fond of relating her experiences from that era. As she grew older, those experiences got more intense, and eventually she was able to recall events from the War of 1812—an impossibility, since it happened many years before her birth.

While failing powers seem to have played a role in this lady's case, most people run into similar problems when trying to sort dimly recollected mental fact from fiction. For instance, many of us can recall early events from our own childhoods. But how many times have we recalled them? How sure are we that the memories recalled are the original memories rather than memories reconstructed and perhaps refurbished in earlier recall episodes? And how sure are we, really, that our recollected memories accurately portray the events as they occurred?

Even so, memories recollected and refurbished from time to time are probably better than ones that are not periodically recalled. That's because nerve connections that are used get strengthened, whereas those that are long idle tend to wither. As a result, the chances of accurately teasing together a complex jigsaw-puzzle memory scattered through the brain after its pieces have remained idle and unassembled for many years is rather slim.

All of this indicates that most childhood memories don't get repressed and retained in the way that Sigmund Freud envisaged. Instead they get widely scattered in the brain and their unused portions simply fade away. But Freud wasn't referring to just any old memories. He was referring to sexual and other memories that were repressed because they were too painful to recall. And while the brain has nothing corresponding to what Freud imagined, we now know that the brain's emergency survival kit could have produced mental effects that inspired Freud to devise his repression theory.

Specifically, as we saw earlier when discussing anxiety, phobic or stress-related memories can get hard-wired into the amygdala; and under conditions of prolonged stress, the hippocampus charged with receiving and distributing memories of ordinary events works poorly. So it is possible for certain stress-related memories to be hidden permanently in the amygdala, even though they have faded from the ordinary memory system or failed to register there in the first place. Such hard-wired memories remain unconscious until triggered by matching events that set off the amygdala's alarm system and cause it to rouse the mind and body to fight or flee.

Freud embroidered his repression theory with all sorts of fanciful ideas about childhood sex urges and early development that are now mostly receiving the benign neglect which they deserve. But he also claimed that repressed memories were responsible for disorders of the mind; and as we have seen, if the amygdala gets hard-wired with hidden memories in ways ill-suited to its owner's needs, that can cause the mind to experience various inappropriate emotional problems—which surface most notably as anxiety disorders. So there was a grain of truth to Freud's repressed memory idea.

We have seen that various therapies can make progress against anxiety disorders. But the chances of actually curing such problems by discovering repressed memories and working through them is poor because the amygdala is so hard-wired. What's more, the amygdala is just one part of the brain. Childhood is just one time when inappropriate conditioning can occur. And such conditioning is just one of various processes responsible for disordering the mind. Thus, as any number of cases ranging from Joyce Harper's to Alfred Ramsey's attest, Freudian or any other abnormal psychology directed mainly at probing repressed childhood memories is too limited to provide a good general picture of mental ills.

Mental states lack fixed boundaries. As noted previously, the mind's state is determined mainly by three variables: the brain's level of activation, source of information, and type of neuromodulation. Of course, multiple neuromodulators and other factors make the actual situation far more complex than it appears in the AIM model. But if we map our three basic variables onto the model's three dimensions, we find that many normal and abnormal states appear continuous with one another. So the model helps to explain why many mental ills (Alice Morrisey's is a case in point) fit so poorly into fixed diagnostic pigeonholes, why they combine with one another in diverse ways, and why they constitute an almost infinitely variable throng of troubles.

More generally, the model shows why mental states are not only multiple and overlapping but dynamically changeable. Nothing is static, as the old "one disease" model would have had us believe. Among other things, this helps to explain why rote prescription based on snap diagnosis often fails to meet even the patient's pharmacologic needs, why both drug prescription and therapy need to be tailored to the case being treated, and why ostensibly distinct disorders like schizophrenia and emotional disorders are often found together.

Both genes and experience are important. Family studies and twin studies have shown that genes play leading roles in autism, schizophrenia, bipolar disorder, and depression. They also have a lot to do with clinical obsessive-compulsive disorder, which shares a genetic heritage with Tourette's syndrome, because a newborn whose family has one of these disorders runs an elevated risk of getting either. And of course, genes have a big say in simple phobias, where an inherited tendency to be afraid of such things as snakes or confined spaces may be too strong to begin with or may be readily escalated into an anxiety disorder by experience.

We have learned less about the roles genes play in conditions classified by DSM-IV as "Axis II" disorders. Nevertheless, common sense and normal experience tell us that personality and character traits can be inherited. "A chip off the old block" says it well. Beyond that, we know genes are involved in the schizotypal personality disorder because people with this disorder run a higher than normal risk of having descendants with this disorder or with schizophrenia. We also suspect that genes are involved in other Axis II ailments like the narcissistic, obsessive-compulsive, antisocial, and borderline personality disorders, because these ailments are generally considered chronic. And if one stays away from the amygdala (which can be hardwired by life experiences), chronic mental problems tend to suggest genetic causes or early damage by biological factors. And since biological factors are not obvious in most personality disorders, suspicion naturally tends to gravitate toward genetics.

But the affected person's experience is important too. Besides hardwiring the amygdala, stressful experiences can pave the way for depression. They can also tilt the brain's chemistry away from thought and reason toward emotion—thus encouraging emotional problems, acting as triggers for a wide range of mental troubles, and causing the affected person to behave in ways that are debilitating or that make the problem worse.

Experience also has a positive side. That's because many counter-productive behavior patterns that have been learned can be unlearned. Events that act as triggers or produce unacceptable stress levels can be identified and targeted for rehabilitation or constructive avoidance. The brain's defenses against built-in problems can be strengthened. And the patient's living arrangements can be changed in ways calculated to improve the odds that future experiences will be positive and beneficial. Therefore, it is important to ensure that the role of experience—both in causing and in treating mental ills—receives the attention it deserves.

For these reasons, the new psychology that we envisage for psychiatry must rest squarely on the strong shoulders of learning theory, the large body of theory dealing with how people learn and come to change their responses. Learning theory is probably the most robust area in all of experimental psychology. While there have been some notable innovations in the application of Pavlovian and Skinnerian principles to psychotherapy, these developments have tended to be idiosyncratic and isolated rather than mainstream and integral. In our view, every psychiatrist should understand the import of B. F. Skinner's claim that behavior is determined by its consequences and should learn to shape adaptive behavior actively through positive reinforcement. At the same time, modern psychiatrists and psychologists must go beyond Skinner's parochial negation of the importance of the brain in setting both the conditions and terms of learning. That is, the brain can no longer be considered a black box. Instead, it must be regarded as the physical embodiment of everything that goes wrong and everything that can be righted in our emotional and social experience.

Mental disorders involve structural brain problems. This follows logically from the fact that the mind arises from the brain. Of course, the term "structural problems" covers a lot of territory—everything from the broad sorts of structural problems seen in schizophrenia to the fine-grained structural problems caused by experiences that hard-wire the amygdala. Even so, the point is worthwhile because it sets brain structures apart from brain chemistry.

That's important, because the structural problems tend to be chronic. For the most part they can't be cured. However, they may not do all that much on their own. Rather, they may render the affected person prone to inappropriate brain chemistry shifts that produce temporary bouts of acute symptoms. This explains why untreated schizophrenic psychosis sometimes resolves itself in a matter of months; why untreated major depression typically resolves itself in less than a year; and why people with such disorders are likely to have relapses—because the chemical imbalances causing acute symptoms come and go, while the underlying structural problems remain.

Structural problems give rise to coordination problems. As we have seen, nearly any mental act worth doing requires lots of coordination. So it seems obvious that structural problems serious enough to cause mental ills will lead to coordination problems. One obvious example is schizophrenia, with its poor coordination between the prefrontal lobes and various other brain structures. But schizophrenia is not alone. Depression commonly involves chronic unbalanced coordination between the amygdala, hippocampus, and prefrontal lobes; bipolar disorder apparently involves failure to properly coordinate the two cerebral hemispheres; clinical obsessive-compulsive disorder seems likely to arise from poor coordination between deep cerebral structures and the cerebral cortex; and anxiety attacks clearly arise from the amygdala's urgent but inappropriate coordination with other areas designed to mobilize the brain. So coordination problems are a hallmark of mental ills.

Brain chemistry mediates disordered states of mind. We have pretty good voluntary control over two of the three AIM factors—how active the brain is and whether we are processing internal or external information. The factor that we cannot control readily is brain chemistry. So it stands to reason that brain chemistry alterations arising from structural and coordination problems bring on most of the unusual mind states seen in mental ills. But most of these alterations are not really all that strange. Indeed, most are well known in other guises. We get many anxiety disorders by applying the chemistry of the conditioned fear response to the wrong cues. We arrive at major depression through the chemistry of grief. And we produce the various psychoses by shifting the brain's chemistry in the direction of REM sleep.

REM sleep deserves special attention in this context, because it offers an alternative to the waking state. It is a second major state of consciousness, one that the mind enters naturally several times a night. What's more, like most of the disordered states of mind, REM sleep favors internal sensory processing and emotion over thought. So going beyond the realization that REM sleep is psychotic, there appears to be a natural parallel between waking consciousness/REM consciousness on the one hand and mental health/mental illness on the other.

That's probably because the mechanisms for getting from one state of consciousness to the other are well established. So if some brain problem alters the waking state's normal chemical balance, a logical way for that balance to shift is toward the other conscious state, thereby favoring mood and/or sensory processing over thought. Of course, this can happen many ways, depending on the nature of an individual's brain and the particular brain chemicals involved; and other established brain mechanisms can also be activated, as noted above in the specific cases of anxiety and depression. The point is not that some particular chemical change must occur. The point is that a number of chemical mechanisms and many possible variations on them are available; so that rather than novel brain mechanisms, it is these familiar mechanisms, activated in the wrong way or at the wrong time, that account for most disordered states of mind.

Proceeding forward, what basic ideas does our neurodynamic approach bring out that relate to treatment? Among others, neurodynamics points up the following:

Psychiatric drugs are not cure-alls. They affect brain chemistry. If used properly they may effectively relieve symptoms of mental ills, usually by shifting altered brain chemistry back toward that found in healthy people and the normal waking state. But in most cases the underlying structural problems will remain, sometimes causing acute symptoms to resurface.

Psychotherapy is no cure-all. It plays a vital role in treating mental ills, but it doesn't cure them. That's because psychodynamic, cognitive, behavioral, interpersonal, and other types of therapy work by altering relatively small numbers of brain connections. This cannot be expected to overcome major structural problems in the brain or even to change hard-wired connections in the amygdala. So like drugs, therapy cannot cure most of the underlying conditions causing mental ills.

Drugs and therapy are mutually reinforcing. Like the coarse and fine adjustments on a microscope, drugs and therapy work far better together than either works alone. Psychiatric drugs, providing crude adjustments that affect the whole brain, can repress the symptoms of many mental ills. But they don't always work, they need monitoring, and they provide no personal relationship or effective guidance for the patient. Therapy, a fine-tuned instrument that engages the patient's conscious mind, can do the three latter things and more. But mental ills are tricky. Just as individual minds vary, so mind problems vary from patient to patient, and straightforward diagnoses are the exception rather than the rule. Therefore, therapy needs to be carefully integrated with drug treatment. That means the person prescribing the drugs should be the same person providing at least some of the therapy and overseeing the patient's care.

Mental ills should be treated like chronic diseases. We know that drugs, therapy, and sometimes the brain alone can resolve many acute problems. But neither nature, time, drugs, nor therapy can remove most of the underlying structural problems altogether; and since a potential for relapse will remain, we should not just be treating mental ills acutely like an attack of strep throat and stopping there. Rather, we should be treating them chronically, checking up on patients with severe ailments from time to time, even when their symptoms are not acute, thereby helping to prevent or contain relapses and promoting continuity of care.

Our ability to treat disorders of the mind is good. Most mental patients will respond favorably to one or another combination of drugs and therapy. Even people like Alfred Ramsey can usually be helped. But that doesn't mean all their problems will vanish. So besides needing capably developed treatment and drug monitoring, they also need a sound therapeutic relationship, guidance, periodic evaluation, and social services tailored to their needs.

If public or private programs providing such services were coordinated well and pursued aggressively on a national scale, they would drastically reduce the human hardship and misery caused at all levels of our society by mental ills. They would make significant inroads into our homeless and drug user populations, because a good many mentally ill people are vulnerable to those evils—which are hard to counter unless the underlying mental illness is being treated. And while such programs would cost money, they would probably cost less than what HMOs, insurance companies, families of the mentally ill, and a broad array of government institutions from jails to homeless shelters are now paying to provide low-quality, underfinanced, and fragmented services.

So why isn't this being done? At the very least, why hasn't failure to do it roused a groundswell of public protest? Two answers seem possible. Either the public doesn't care about mental illness; or else the public doesn't understand what mental illness is and what good psychiatric care can do. We think it's the latter. We think there is widespread misunderstanding on several counts. For one thing, we think that many people still associate psychiatric therapy with psychoanalysis, and they're worried about being asked to pay for lots of questionable therapy sessions. For this reason, they fail to see the need for close coordination of treatment or the vital role that psychotherapy plays in dealing with mental ills, and so they feel all right about limiting the role of psychiatrists to prescribing drugs.

We also think that many people in our high-tech era have been oversold on biomedicine. So they regard psychiatric drugs as wonder drugs that can cure mental ills in a clean straightforward way—like hitting a Little League home run. They don't see that the drugs are limited, mental ills are quirky, and each patient is different. So providing treatment, including drug prescriptions, is not like hitting a Little League home run. It is more like playing Major League baseball and trying to hit a well-hurled knuckleball with a warped bat. As the sad case of Alice Morrisey demonstrates, that makes close coordination of drug prescription, psychotherapy, and other services essential.

Then too, a lot of people still think that mental ills are phantasmagoric mysteries—more like voodoo magic than rheumatoid arthritis. The old Freudian mystique promoted that. And biomedicine tends to provide drugs without much explanation, leaving an aura of mystery still hanging. For this reason, people need to know what brain science has discovered.

They need to know that virtually all psychiatric ills arise from neurodynamic disorders of the brain; and they need to know that these disorders, despite their sometimes dramatic symptoms, are no more mysterious than gallstones or arthritis.

Finally, many people have trouble registering the fact that most mental ills have a chronic side. Distracted by the spectacular ups and downs of many ailments, they have focused on acute symptoms as though they were the whole story. This has let the HMOs and insurance companies off the hook, allowing them to cover only treatment of acute conditions. In millions of cases, this has compromised the continuity of care, causing patients to be lost to treatment and promoting homelessness, misery, imprisonment, drug abuse, and crime; so it seems clear that this is another misunderstanding that needs to be cleared up.

These misunderstandings show how badly we need neurodynamics. For today's psychiatry is trying to work in a psychology vacuum. Much or even most of Freud's psychology has been discredited. But nothing major has replaced it. So most psychiatry residents and psychology graduates are still receiving their therapeutic training from humanities-oriented psychoanalysts steeped in Freudian thought. That has left most therapy effectively isolated from biomedicine, deprived of scientific roots, and burdened with a publicly bankrupt rationale.

Of course, there are all kinds of other psychologies ranging from respectable to kooky. But none has replaced Freud's pitifully outdated guesswork. That's partly because the Freudians are well entrenched in their psychoanalytic institutes, and partly because psychodynamic therapy can be marginally effective even without a valid rationale. But the main reason for this failure is that brain science has been learning volumes about how the brain and mind actually work. That's *really* exciting. So while brain science keeps churning out new knowledge, no freshly minted top-down psychology can hope to compete, catch the public eye, and replace the old Freudian mystique.

But something can. That something is neurodynamics. For neurodynamics does not compete with brain science. It embraces brain science. It uses brain science to provide a sound foundation for both psychotherapy and biomedicine, and in so doing it seeks to heal the breach between them. What's more, neurodynamics is an *approach* to psychology, a method rather than a doctrine. So besides integrating current brain science discoveries, it stands ready to embrace future discoveries.

In a way, this revolutionary neurodynamic approach represents little more than an idea whose time has come. For brain science has been making strong progress, and its emerging discoveries about the mind will be incorporated into both psychology and the treatment of mental ills sooner or later. But sooner is better than later—because later can be *much later* if no one acts, and because our past failure to fill the psychology vacuum is now damaging psychiatry, devaluing therapy, interfering with proper care of mental ills, and encouraging a host of public misunderstandings that stand in the way of sensible and strong corrective action.

So in our view neurodynamics is not an option. It is not a frill. It should lead the charge against today's bizarre and mismanaged treatment of mental ills. And it should be actively pursued by a broad array of psychoanalysts, cognitive scientists, psychiatrists, psychologists, therapists, and others—indeed, by everyone interested in seeing psychiatry revived and assigned its proper role. That's not just because neurodynamics can help heal psychiatry's split personality, promote coordinated services, and restore therapy to its rightful place. It is because at heart our current difficulty in rationally applying the findings of brain science to the treatment of mental ills arises from a lack of public confidence in psychiatry. And sound application of neurodynamics offers the best possible way of turning that around, drawing attention to our cause, and restoring the public's social and political willingness to act.

$$\boxed{13}$$

The Road to Reform:
What Can Be Done Now?

Suffering, and nothing else, will implant that sentiment of
responsibility which is the first step to reform.

—James Bryce,
The American Commonwealth, 1888

The time for reforming psychiatry has arrived. But transformations like this don't happen on their own. They need help. So besides enlisting help from everyone who can give it, building a scientific foundation, and encouraging the public to see things as they are, we need to assess what must be done to alter the system in ways that will make the most obviously needed changes come about.

Many changes seem both desirable and needed, for psychiatry truly is in crisis. As we have seen, public mental hospitals are still closing and turning their work over to the private sector. But private psychiatric units are also closing because they cannot make ends meet. Hard-pressed hospitals, HMOs, and insurance companies are turning psychiatrists into short-order diagnosticians and pill-pushers. Most domestic medical school graduates now shun psychiatry. Many current psychiatric residents are foreigners whose medical qualifications and command of English are limited. Jails and homeless shelters are trying to do what the dwindling supply of mental hospitals once did. Coordination of care for the mentally ill is poor. And public discontent with the resulting patterns of neglect, misery, and violence is on the rise.

Despite this rising discontent, today's reformers have only limited resources. They are unlikely to get more while public confidence in psychiatry remains uncertain and current trends toward privatizing hospitals persist. So they should not hesitate to ask for more resources. But they should also husband what they have. They should bow to strong prevailing trends rather than resist them. And they should respond creatively to current needs in ways that raise public confidence in what psychiatry can do.

For instance, psychiatric resident training and education methods need an overhaul. But the prevailing methods are well entrenched, so we might best start with modest changes in undergraduate and medical school education and recruitment. In a similar vein, we desperately need more hospital beds but seem unlikely to get them; so we should focus instead on reaching the neglected mentally ill where they are, most notably in homeless shelters and in prisons. We should also try to improve the continuity of care, reduce heavy court costs associated with psychiatric cases, and encourage the new trend toward outpatient commitment.

Each of these points deserves a closer look.

Regarding education, most current residency training is apprenticeship training. Lectures get relatively little attention, and brain science lectures get hardly any—in part because private hospitals can charge substantial fees for contacts between residents and patients, and so the residents are under steady pressure to maximize their work with patients. As a result, even the pharmacology training tends to be "top-down" medical apprenticeship training in diagnosis, hospital procedures, and drug prescription. The "bottom-up" brain science basis of mental ills and psychiatric drugs gets short shrift. And of course, with no common foundation, medical pharmacology training and humanistic psychotherapy training remain like Paolo and Francesca in Dante's *Inferno*: doomed to cycle endlessly in a fruitless desire for one another.

It's not hard to see the futility of trying to break this mold with half-measures like providing lectures on relevant brain science. Since the residents' time is valuable, one would have to pay the hospital for the privilege of doing this, or else the hospital would have to bear the cost; the residents are under heavy pressure to see patients rather than attend lectures; and many humanists who teach psychodynamic psychotherapy have a negative attitude toward science. So the costs would be high, interest might be low, and attendance would run from sparse to nil.

These days, the people who are learning brain science and psychology *en masse* are undergraduates. There's lots of interest in these subjects in college, so lots of undergraduates emerging from our better universities have a firm grounding in both. What they generally lack is a neurodynamic understanding of how the two things fit together. So they have no firm sense of how to connect brain science with psychology; and they also lack an understanding of psychopathology, the discipline that relates psychology to mental ills.

Many of these kids get to medical school. There they find psychiatry afflicted by this schismatic setup in which therapy is divorced from biomedicine. They also come to see that many practicing psychiatrists are pressed into pill-pushing roles with minimal exposure to their patients. And so, for one or both reasons, most of them opt out of psychiatry and head on into family practice, where intimate contact with patients is assured, or into neurology, where the brain science base is well established and the future seems more promising. That's all right for family practice and neurology, which of course should not be neglected, but it clearly leaves psychiatry in the lurch.

If we want to change this pattern, and it surely needs changing, one should obviously start at the bottom with the undergraduates. For instance, things could be improved by developing undergraduate psychology and brain science courses that take a neurodynamic approach. That would encourage students with interest and ability in both areas to bridge the gap between them. It would also nurture a pool of students with more than a passing interest in both psychology and brain science—potential future psychologists and psychiatrists dedicated to treating the brain and mind together.

Beyond that, courses with this neurodynamic orientation should be offered in medical school, where they could help to heal the split between psychotherapy and brain science. Such courses could combine productively with the medical schools' clinical work that is now so much in vogue—because they could help students to get a panoramic view of patients' psychiatric needs and the brain basis of mental ills simultaneously. Such training could provide the psychopathologic knowledge of mental ills that is usually missing from undergraduate education, permitting the medical schools to serve psychiatry in the same way that they now commonly serve neurology and most other specialized fields of medicine.

Finally, lectures with this orientation could be used at teaching hospitals and elsewhere to provide continuing education. Admittedly, lectures play a relatively small role in resident training. But if colleges and medical schools are providing a neurodynamic foundation, then in time a fair share of the teaching hospital instructors will find themselves comfortable enough with neurodynamics to apply it, and neurodynamic lectures can complement their work.

Many teaching hospitals—all those that do interesting or compelling work—could also improve their recruitment in various ways. Suppose, for instance, that Mass Mental or some other teaching hospital begins by offering fellowships for undergraduates to come get an idea of what is happening. This recruitment-oriented program could continue in medical school with the same or different students and could provide some kind of structure—including special orientation meetings, a limited curriculum, search opportunities for mentoring relationships, social gatherings, and so on. It would really be a recruitment and demonstration program for psychiatry. If successful, it could later apply to a foundation or the federal government for funding to support its activities, and could also serve as a model for recruitment programs that could be adopted elsewhere.

Of course, it makes no sense to train and recruit qualified psychiatrists if they can't have proper access to their patients. That means we need to consider another facet of psychiatry's dilemma, which is the vanishing hospital beds problem. As noted earlier, the number of public mental hospital beds in the United States fell from around 558,000 in 1955 to less than 60,000 at the end of the millennium. This drop was somewhat sharper in Massachusetts, where the numbers went from some 25,000 beds in 1955 to less than 2000 now. No matter how much these missing beds are needed, current trends suggest they aren't coming back anytime soon. So the obvious answer is to reach the severely afflicted mentally ill outside the hospital, where they live, rather than inside it.

In the case of homeless people, that means working with a wide range of public and private agencies trying to help the homeless. And while this requires a certain amount of diplomacy and coordinating skill, it also takes advantage of several favorable circumstances. For one thing, as the inpatient population shrinks, residents and other staff members have fewer inpatients to work with; so rather than cutting back on residency training and other programs, it makes sense to use the available pool of skill outside the walls. And clearly, services externalized this way can avoid many of the mental hospital's normal in-house overhead costs, making these services relatively cheap.

Moving in this direction really amounts to nothing more than providing some of the community services touted in the 1960s when the mental hospitals were decanted—services that were promised but generally not delivered. This helps to explain why such action can be so productive—because these community services have never been effectively provided, while the old abolished services have never been restored; and so hordes of neglected, drifting, homeless people with severe mental ills have become an established part of urban life.

That doesn't mean that providing psychiatric liaison or treatment where the homeless live is an ideal arrangement. Many of the mentally ill homeless lack insight into their illness, and so have little or no interest in being treated. Unfortunately, this lack of insight tends to rise with the severity of the ailment, being especially pronounced in people with bipolar disorder and schizophrenia. What's more, many homeless people with severe mental ills compound their problems with drug or alcohol abuse, making shelter-based treatment more difficult than it would be otherwise.

Even so, a lot can be done under adverse circumstances. Many current tales in the psychiatric realm attest to this, but few do so better than the recent history of the Massachusetts Mental Health Center. When we left that institution in Chapter 4 it had fallen on hard times. Its staff had been cut, its budget was shrinking, and its physical plant was deteriorating. So while its teaching and research staff was still first-rate, Mass Mental had entered a twilight zone of neglect and slow decline. Finally, in the early 1990s, its in-patient unit was shifted over to Boston's private Deaconess Hospital. At this point, Mass Mental's psychiatric residents had no inpatients to work with; state government officials began eyeing Mass Mental's valuable urban land site; and some began sharpening long knives with an eye to Mass Men-tal's ultimate demise.

But Mass Mental wasn't finished. Indeed, until it lost its inpatients it was still at the top of just about everyone's list for training in public psychiatry. And so, while little could be done about its long-standing budget and main-tenance problems, the immediate question was what to do about its resi-dents.

This problem was not quite so insoluble as it appears. For one thing, Mass Mental no longer did all the residency training in its area. It had joined forces with other nearby hospitals—Beth Israel, the Brigham and Women's, and the Deaconess—to provide joint training for psychiatric resi-dents in the so-called Longwood Area.

Nor was Mass Mental bereft of patients. There were still plenty of outpa-tients, including some 400 living in various sorts of community residences connected to Mass Mental and scattered about the center's large urban catchment area. Also, it still had a 50-patient day hospital set up by Lester Grinspoon in the 1950s to provide homeless and other mental patients with a 60- to 90-day program of treatment, services, and rehabilitation; and it still had an overnight shelter, the "Fenwood Inn," that it had established to accommodate homeless patients, including those at the day hospital. Even so, these arrangements did not provide an ideal base for training resi-dents, and so it was decided that residents doing their Mass Mental rotation would be sent out to treat the homeless in other settings.

The person largely responsible for doing this was Kenneth Duckworth, a promising young psychiatrist bent on community service who had com-pleted his own residency at Mass Mental a few years earlier, and who found himself in charge of the residency program. Duckworth started out by contacting Health Care for the Homeless, a nonprofit outfit that provided medical care (but not mental health care) for homeless people in Boston and other cities. He asked how their Boston organization would like the services of a group of crackerjack Harvard psychiatric trainees free of charge, the only catch being that Duckworth and others would need to come along and train them.

Health Care for the Homeless knew very well that many homeless people had mental problems, and after thirteen years of experience it was ready to take the plunge into mental health; so it accepted Duckworth's offer. The result was a strong partnership that turned Mass Mental's treatment mission away from its aging buildings and toward community-oriented service to the homeless. It also provided a parking place, what became known as the "homeless rotation," for the Longwood Area's third-year residents. Beyond that it broke new ground, being the nation's first psychiatric residency rotation devoted solely to working with the homeless. It provided a healthy challenge to the residents, because homeless people tend to have tough psychiatric problems. And since most residents really wanted to help the indigent and homeless, the program was wildly successful, becoming the single most popular rotation in the Longwood residency.

Encouraged by success, Duckworth proceeded to spread out. He got residents in his rotation to provide consultative services for other groups—among them a women's homeless shelter, a state homeless outreach team that pounded the streets of Boston to offer free sandwiches and services, and an "old girls' network" of gray-haired social workers who sought to provide housing for the homeless through an organization called the Committee to End Elder Homelessness. He also arranged for two well-qualified psychiatrists to do liaison work with Boston's enormous homeless shelter, the Pine Street Inn. And he helped set up a user-friendly walk-in clinic at Mass Mental, where any homeless person in the city could go three days a week around lunchtime to receive a free meal and professional attention.

As previously noted, Mass Mental's community approach to mental health goes way back in its history. But its strong recent initiative to help the homeless provides a good example of adaptation under adverse circumstances—something badly needed in psychiatry these days—and shows what can be done with scant resources.

Of course, it's all very well to move services out to where patients live; but if we do that, we have to improve coordination between the complex of agencies providing treatment. This means that to make real progress we must grapple with another of psychiatry's problems—the urgent need to improve the continuity of care. Clearly, a whole series of issues needs to be addressed here—including the patient's right to privacy, the willingness of different agencies to cooperate in any given area, and the costs involved in keeping and transferring records. But while we have been living in the computer age for many years, it seems evident that coordinated monitoring of uninsured, indigent, and homeless patients is very poor. So it seems likely that record-keeping and patient treatment could be better coordinated in most places, and that this would benefit the patients.

Of course, this need for continuity applies within treatment centers as well as between them. So if a patient goes bouncing into a mental hospital or treatment center for the second or even the umpteenth time, treatment should be provided by people who have treated the patient previously and who have access to that patient's prior records. This may seem burdensome for crowded or fiscally troubled hospitals and other centers that have scant time to spare. But revolving doors have entrances as well as exits; overburdened hospitals and other centers do not want to see the patients coming back again if they can help it; and providing a modicum of continuity offers the best chance of preventing that, by preventing repetition of measures that haven't worked and encouraging adoption of ones that have.

Thus, even private facilities battered by market forces and insurers can gain from improved continuity. But the public sector has far more to gain. Not just mental hospitals but also general hospitals, homeless shelters, jails, and ultimately the taxpayers have a large stake in keeping mentally ill people from being lost to treatment. That's because being lost to treatment makes patients more vulnerable to suffering an acute attack of illness, becoming a danger to themselves or others, and landing in yet another public institution. So the public sector has good reason to want coordinated treatment, to insist on coordination with the private sector, and to recognize that when someone with a severe ailment like schizophrenia or bipolar disorder fails to appear on schedule, the time has indeed come, as NIMH Director Steve Hyman points out, to find that person and determine what is happening.

Obviously, not all neglected people with severe mental ills can be found in homeless shelters or the streets, because many are in prison. And it does no good to establish fine continuity of care through services reaching homeless people if they get arrested and the thread of continuity is snapped. So besides revamping education and recruitment, reaching out to the mentally ill homeless, and pushing for improved continuity, we also need to reach the mentally ill who are in prison. As already noted, somewhere between 8 and 16 percent of our roughly 2 million jail and prison inmates appear to suffer from severe mental ills (schizophrenia, bipolar disorder, major depression). That means there are far more mentally ill people in prison than in mental hospitals, receiving treatment ranging from marginal to nil.

Even worse than the uncertain treatment is the setting. For prisons are mainly dedicated to handling and sometimes reforming criminals by imposing lots of rules. Prisoners with severe mental ills are not in a good position to obey such rules, and of course they are easy prey for hardened criminals. So they get punished, they get victimized, and they gum up the works of the prison system, demonstrating the merit of the statement that "the bad and the mad just don't mix," meaning not that they can't but that they shouldn't.

This picture of prison life seems at least as misguided as anything found in the pages of Samuel Butler's *Erewhon*, where criminals are given medicine and sick people go to jail. So what can we do about it? We can blame public misunderstanding, our poor misguided laws, and our network of sometimes hard-hearted bureaucrats, hospital administrators, and insurers bent on shrinking psychiatric units and kicking psychotic mental patients out into situations where they are likely to get arrested and tossed in jail. We can get quite irate about all this. Or we can see that trying to break the system is like trying to break down the Great Wall of China with our hands; that *Erewhon*-like scenarios can help as well as hurt the mentally ill; and that if we seek to work within our crazy system, reasonably good answers can be found. For as Dr. Erik Roskes noted recently in the journal *Psychiatric Services*, "Mentally ill [imprisoned] offenders can be treated. It is time for psychiatry and other mental health disciplines to include these difficult-to-treat but needy clients in their mission."[1]

A good way to start is to hark back to the much-maligned mental hospitals of the 1950s. The main charge leveled at these institutions then was that they were warehousing and mistreating the mentally ill. Fair enough. Sometimes they were. But isn't that what today's prisons are doing with a vengeance? What's more, today we have the tools and knowledge to help most mentally ill people. So if most people with severe mental ills can't get into mental hospitals because of poor insight, laws against involuntary commitment, and lack of space; and if many of these people are winding up involuntarily in prisons, why not bring key elements of mental hospital treatment to the prisons?

The first logical step in this direction would simply be to separate the mad from the bad. That would need to be done carefully.* But those in charge would know the prisoners and have their histories. Furthermore, we know a lot more about mental ills than we did in the days when someone could get into a mental hospital simply by claiming they heard voices. It seems unlikely that many regular prisoners would wish to be classified as mentally ill and confined with "crazy" people in a separate wing within the prison walls. And it also seems unlikely that such a ruse would work for very long. So there is good reason to think that such a separation would succeed; and it also seems likely, even if nothing else were done, that this would benefit all parties by preventing a good deal of disciplinary breakdown, victimization, and abuse.

---

*The main effort should be directed at identifying prisoners with severe mental ills (mostly depression, bipolar disorder, mania, schizophrenia, and admixtures of these). The American Psychiatry Association's 1994 (fourth) edition of *The Diagnostic and Statistical Manual of Mental Disorders* (DSM-IV) includes these within a large category that it calls "Axis I" disorders. Another group of psychiatric problems, referred to in the manual as "Axis II" disorders, tends to be fuzzier, less obviously debilitating, and harder to diagnose. Also, Axis II disorders include among their number "antisocial personality disorder," which corresponds pretty much to the stereotypical male criminal personality type. So selecting people for inclusion in the "mad" group because they have this particular Axis II disorder would be self-defeating.

Further steps could be directed at changing conditions within the psychiatric wing to conform with the real needs of mentally ill inmates. Some of this could be done in a primitive way merely by retraining members of the regular prison staff in certain basic principles of mental hospital care. But most prison staffers tend to emphasize the need for discipline and minimize psychiatric needs regardless of retraining. What's more, our country seems very ready to jail people these days; lots of people are in prison; and so the prison's psychiatric wing would house only a small share of the total prison population. As a result, no matter how this wing were staffed, few regular prison employees would be displaced, and the staffing changes would not greatly affect the prison budget. For these reasons among others, it would seem reasonable to step up the amount of psychiatric treatment provided for inmates in the psychiatric wing, and to staff that wing mainly with people experienced in coping with confined mentally ill patients.

If this were done well, in a way that actually removed the mad from the bad and provided them with rational treatment, it would notably improve the quality of life, not just for the mentally ill inmates but for everyone working or confined within the prison. It would not change existing release practices, nor would it alter the sentences of those confined. But it would, in effect, bring the mental hospital to the prison. It would provide much more effective care for mentally ill inmates than they get now—something with potential long-term social benefits. And it would pave the way for better coordination with outside agencies when mentally ill inmates are released.

This last is no small thing, because far more mentally ill people are on probation than in prison. Though figures vary, a recent U.S. Justice Department survey indicates there were about 550,000 mentally ill probationers in mid–1998. Since most probation workers are concerned mainly with disciplinary matters (whether the parolees are meeting their conditions of parole), they are in a poor position to give top priority to mental health. But we know that disproportionate numbers of released prisoners with severe mental ills tend to become indigent or homeless after their release. So clearly, if long-denied community outreach services to the mentally ill homeless and impoverished are expanded, mental health authorities will have a golden opportunity to close the circle by linking services reaching the poor and homeless mentally ill with the prison system's mental health records in a way calculated not to limit freedom but to vastly improve the continuity of care.

This brings us to outpatient commitment. Nearly all states now have laws permitting "commitment" of mental patients to treatment outside of mental hospitals. In theory, the process works like this: If a court finds that a mentally ill person meets the criteria for outpatient commitment (usually posing a danger to oneself or others) it can order that person to follow a treatment program. The program, which should be specifically tailored to the individual, can include such things as taking a particular course of medication, visiting a facility to permit monitoring, attending individual or group therapy sessions, and participating in educational, vocational, or substance abuse programs.

While mandatory outpatient treatment seems logical at a time when the number of inpatient beds is pitifully small and shrinking, it does have certain drawbacks. For one, it must be ordered by a court, so it involves court costs and absorbs the time that various high-paid professionals—including psychiatrists and attorneys—need to prepare a case. Also, in many states (a majority as of this writing) the only people subject to mandatory outpatient treatment are those whose mental ills make them dangerous to themselves or others. But those who meet this high standard of peril—imagine people who are acutely psychotic, demonstrably suicidal, or irrational enough to walk in front of traffic—are unlikely to benefit from mandatory outpatient treatment, because they are unlikely to obey the judge's orders. They are also prime candidates for the public mental hospitals' shrinking supply of available beds. So it is hardly surprising that many states with tightly restricted "mandatory outpatient treatment" statutes have tended to bypass this outpatient option in favor of hospitalization.

But this pattern may be changing. Increasing numbers of states now have laws that permit involuntary outpatient treatment on a less restrictive basis. Most prominently, New Yorkers repelled by the well-publicized psychotic subway killing of Kendra Webdale and other troubles have pushed through something dubbed "Kendra's Law." The state statute, which went into effect in November 1999, allows the New York courts to order involuntary outpatient care (but not forced medication)* for people incapable of giving their informed consent to treatment. The new law produced only a handful of commitments during its first few months in force—perhaps because drugs play a big part in treatment, many cases are settled out of court, community services may not have been available, and courts may have found the measure unfamiliar. Of course, there have been strident objections to this and similar laws by civil libertarians who put the abstract concept of patients' rights ahead of effective patient treatment or public welfare. Even so, the growing movement toward less restrictive outpatient commitment suggests that many people are fed up with blatant public neglect of the mentally ill and are looking for some sort of rational balance.

Outpatient commitment alone cannot produce that balance. Various studies, including a long-term pilot program at New York's Bellevue Hospital, have found that outpatient commitment programs can reduce the hospital revolving door syndrome by lowering the number of readmissions. But getting a court order for outpatient commitment is expensive. Relatively few cases seem likely to be decided by a judge in this manner. And it's not precisely clear how much program benefit derives from the court order and how much from the intensive community services that go with it.

What does seem crystal clear is that outpatient commitment without strong outpatient services is worthless. The formal "Resource Document on Mandatory Outpatient Treatment" produced by Joan Gerbasi and others for the American Psychiatric Association's Council on Psychiatry and Law says this in a more positive, upbeat manner: "It appears that mandatory outpatient treatment can be a useful tool in the effort to treat chronically mentally ill patients with documented histories of relapse and rehospitalization. It is important to emphasize, however, that all programs of

---

*Involuntary outpatient care without forced medication seems a little odd, considering the major role that drugs play in subduing severe disorders. It is true that drugs can be difficult or impossible to administer involuntarily in an outpatient setting, and so this approach has both pros and cons. But the real point is not the pros and cons of this or any other approach to outpatient commitment. The real point is that these laws arise from growing popular impatience with current troubles and reveal mounting public willingness to act.

mandatory outpatient treatment must include intensive, individualized out-patient services."[2]

Sounding a positive, upbeat note is important, for this new direction in law could signal something more than just well-meaning legislation in a vacuum. It could signal renewed public willingness to set up and support mental health community outreach services, so long as those community services are useful and cost-effective. In other words, an opportunity may be arising at long last to recover from the community outreach disasters of the 1960s and to move ahead in a manner that is at once less ambitious and more creative to fashion community outreach services that work.

Looking to the future, it seems relevant that forty years have passed since our nation last assessed its mental health system with a view to changing existing practices and laws. Clearly, in view of the current picture, the time is ripe—indeed overripe—for this sort of review. Of course, other considerations come into play here, such as whether the federal government favors such action and whether the general public supports it. While these factors don't affect the urgent need for such an assessment, they make its timing critically important.

We have no crystal ball, so we have no way of telling just when the proper time might come. But we think that it will come; and if psychiatry shows serious signs of putting its own house in order, this time could come relatively soon. Therefore, it seems reasonable to suggest a few general matters that anyone involved in conducting such a review and making recommendations should consider. These matters are as follows:

First, any mental health care system worthy of the name needs to deal with substantial numbers of people who are chronically or periodically psychotic. While psychotic, such people do not necessarily have good insight into their condition; so they tend to fall through the cracks, and any decentralized treatment system finds them hard to handle. But large numbers of people are involved; so failure to handle them raises the prevailing national levels of misery, suicide, alcoholism, drug abuse, violence, and incarceration. It also incurs major social costs—including the costs of running an inefficient treatment system; the costs of additional court, prison, and homeless shelter services; and the costs of dealing with neglected or poorly

FIGURE 13.1   Boston's Massachusetts Mental Health Center is still strategically situated. Conceivably, it could play a strong role in reforming today's psychiatry—but only if torn ties between its academic and patient care missions are mended and its aged and neglected physical plant receives a long-needed overhaul.

treated patients who, if they were properly treated, would stand a far better chance of staying out of trouble or even remaining gainfully employed.

Perhaps the best way to dispel this miasma of poorly coordinated care would be to set up a hundred or so regional mental health centers for diagnosis, treatment, and research (Figure 13.1). Ideally, each center would serve roughly the same number of inhabitants, and each would act as the hub for a network of strong, close-knit community outreach services—including residences, group homes, rehabilitation facilities, and neighborhood health clinics. These centers would be staffed and equipped to ensure that incoming patients receive high-quality evaluation. Such initial evaluation—together with other possible services including triage, family education, and long-range planning—would be conducted on an inpatient basis. Of course, because so many facilities capable of providing inpatient care have closed their doors or been shifted to other tasks, significant investments would be needed to build, convert, or revive such facilities and also to recruit capable professionals and health workers to staff them.

Besides needing money to get established, these centers and their associated community support services would require operating funds. But in return they would provide major social benefits. In fact, they would provide such extensive social benefits that we need to think clearly about the balance between the public and private sectors in this area and about the regrettably chronic nature of mental ills. For the sad truth is that mental disorders cannot be treated without cost; and indeed, no private sector insurer or HMO can hope to deal with them adequately on its own in a way that produces a desirable-looking bottom line. Therefore, we need to develop a realistic public mental health program that benefits both society and medical insurers—by either subsidizing or eliminating the mental health role of the insurers.

Obviously, such a program would need to bridge the gap between federal and state government. One way to do that would be through federal "capitation" grants to the states. Such grants could be given to states that agree to match them with local tax revenues, as well as to private facilities that request them, and their size could be based on the numbers of people in the regions served who are found by standardized epidemiologic studies to have certain mental ills. Provision of funds would depend on the ability of both the state and private facilities to meet carefully devised state and federal standards for inpatient and community services.

Would these new centers threaten to revive the evils of the old state mental hospital system? That seems unlikely. To begin with, the centers proposed here and the old state hospital system are vastly different—probably as different as David and Goliath. True, the state hospitals started with an enlightened aim and then found themselves flooded with patients. But the relatively small centers proposed here would provide only short-term care limited to something like two months—enough to see that the patient was responding to drugs and therapy and was no longer psychotic—and would then assign the patient to community services. The center would bear responsibility for making periodic follow-up visits to assess compliance with its recommendations and the adequacy of treatment, but the main thrust would be directed at outpatient care.

Of course, even this short-term care within the centers would require noteworthy modification of state laws, most of which prohibit any involuntary commitment or treatment without a guardianship arrangement or a court order indicating that the patient is a danger to himself or others. As things stand now, the chances of modifying such laws are slim. But if psychiatry undertook to reform itself and public enthusiasm rose, even civil libertarians might come to see that a short-term course of involuntary commitment and treatment is preferable to a long-term course of neglect, homelessness, and involuntary incarceration.

What then of that small minority of center patients who fail to respond to treatment and remain psychotic? Barring some more fundamental changes in existing law, there would be two choices. Those who assented (or were deemed dangerous or had some guardianship arrangement) could be assigned to state hospitals designed to care for them until the psychotic state abated. The rest would be released in accordance with the law. Many would follow the familiar route to homeless shelters or to prisons and perhaps to psychiatric care in one or another of those places. It should be noted, however, that the number traveling this route would be small compared to the number pushed that way by today's confused and fragmented system; and if efforts to improve psychiatric services for those in prisons and homeless shelters prove successful, even these people would stand some chance of getting proper care.

So how might we ensure that our network of centers, community services, and hospitals is well staffed and coordinated? Again, we have no crystal ball. But the same forces (brain science and public support) needed to pave the way for our proposed national assessment and sound funding might well create a stream of enthusiastic and qualified recruits for psychiatry and might also improve the ability of therapists, psychologists, and psychiatrists to work together.

Of course, it would be foolish to take this for granted, or to assume that learning obtained in college or medical school would necessarily be updated. Rather, education should be built into the system, so that in order to qualify for financial support each center would need to demonstrate that it had mandatory ongoing education programs for its own personnel and those of the associated community services—including all psychiatrists on the regular staff, psychiatric residents, medical students, psychologists, psychology interns, counselors, nurses, nursing aides, and social workers. At the upper staff levels, these programs—essentially continuing education programs designed to supplement apprenticeship training—would devote special attention to the brain sciences (including cognitive neuroscience), to developmental psychology, and also to perceived gaps in pharmacologic and therapeutic knowledge.

This push for education should also extend further. That's because patients as well as their caretakers, friends, and family members sometimes lack effective access to accurate information about the mental ills that they confront, and sometimes individual health workers are too poorly informed or too busy to provide it. The emergence of patient support and advocacy groups in recent decades has improved things quite a lot. Even so, each center should ensure that appropriate information is provided—and also that information clearinghouses are established—for the purpose of informing patients, their caretakers, and others immediately involved. The principal aim of all this, besides answering relevant questions and providing reassurance, would be to increase treatment compliance and strengthen the patient–caretaker alliance.

On a more fundamental tack, one attractive way for a center to promote both education and research would be to link itself with a university, preferably one with a medical school and one that is turning out significant numbers of graduates well-versed in brain science, psychology, and the neurodynamic bridge between them. That would do more than provide access to the university's library and staff. It would make the center and its associated services a logical place for university and medical school graduates to work. It would also attract scholars seeking to do teaching and research and would set up a cross-fertilization process with the university that would act as a general stimulus for accomplishment—something that could markedly enhance the center's image and performance.

Turning specifically to research, psychiatry today faces a critical lack of access to sophisticated research tools and is also experiencing a marked drop in clinical research. To help reverse these trends, each proposed center should be equipped with modern research tools, including the magnetic resonance imaging (MRI) scanners needed to effectively observe the brain at work. Because this equipment should permit advanced research, it should ideally be at the leading edge of new technology, or in any event should not lag far behind.

Of course, tools by themselves are useless. So each scanner should have its own team of technical operators. More important, each center should have its own group of faculty members collectively in command of various scientific disciplines—including genetics, molecular biology, neurophysiology, neuropharmacology, and neuropsychology—and clinically experienced in dealing with most types of severe mental ills. These faculty members should be equipped by their experience and reputation to compete for the grants needed to fund specific research projects conducted at the center.

Such research would typically be carried out with patients who are assigned to the center, associated community outreach services, or associated hospitals. Naturally, this would involve important moral, legal, and financial obligations to each patient that should be spelled out clearly in advance, in addition to having each project's plans subject to rigorous review by appropriate oversight bodies. The fact that patients would spend relatively little time at the center would prove an inconvenience to long-term projects; but continuity could be maintained through the community outreach services or hospitals, and explaining the pros and cons of participation clearly when each subject's advance voluntary consent was obtained could reduce the risk that the subject would later have a change of heart.

At present, brain science is advancing fast. So at any time in the future when suggestions like these are being considered, psychiatry will need to take full advantage of what has already happened. This suggests that investigators at the centers should focus first on applied research with practical implications that can yield timely therapeutic benefits.

Researchers dedicated to helping people will find that quite exciting. But even so, we should not lose sight of something else. Namely, we really are coming to understand the human mind. For the first time, thanks to the advances of brain science, we are starting to see how the master magician does his work. We are starting to unravel the secrets of sleep and abnormal states of mind. And we are starting to discover the true nature of mental ills.

What this means is that our drive to reform psychiatry and press forward with psychiatric research on mental ills relates to something more. It relates to humanity's scientific quest to further reveal the secrets of the mind. This exploration of humanity's inner space is not anywhere near as dramatic or expensive as our parallel planetary exploration of outer space. But it is at least as successful; and it has an array of implications for everything from mental ills to computers to general human knowledge that shows promise of vastly exceeding the significance of anything being learned from the planets.

Right now our principal goal, one we know we can achieve, is to improve the lot of multitudes in urgent need of help. But in addition, psychiatric research seems poised to do more by contributing to fresh discovery—much as the global navigator Ferdinand Magellan did more than improve the lot of Spain through his search for a new route to the Indies. Those joining our quest to reform psychiatry should draw inspiration from this fact.

As anyone can see, these ideas are directed mainly at the future. Right now psychiatry is in a parlous state, and significant actions taken to improve things now must probably be taken more or less at current funding levels. Still, it is worth keeping an eye on the future. That's because any progress achieved today through enlightened action will be but a pale forerunner of what stands to be accomplished if psychiatry truly embraces its brain science base and awakes the general public to full realization of what can in fact be done.

For we have passed the time when we should be warehousing the mentally ill in asylums or in prisons. We have passed the time for putting everyone on couches and pretending to knowledge we don't have. And we have also passed the time for alleging that serious mental disorders can be cured with a bunch of pills. As we have seen, science has learned a lot about the brain. This knowledge has shed a powerful light on mental ills that besides pushing back the darkness has revealed much about how to combine medication, therapy, monitoring, and other features of sound treatment.

The situation is not simple, just as the situation that faced the health pioneers who launched modern medicine a century ago wasn't simple. So we cannot afford the luxury of pretending that mental ills can be treated in a generally enlightened manner without brain science knowledge, by ill-qualified pill-pushers, without hospital beds, in homeless shelters, or in prisons. We can improve things quite a bit at current funding levels. But we have the knowledge to improve the lot of millions and push back mountains of social misery and waste. That cannot be done without more funds; more funds depend on regeneration of public confidence in psychiatry; and regeneration of public confidence demands drastic strengthening of the bridge between psychiatric practice and humanity's growing scientific knowledge of the brain.

Right now we see some hopeful signs. We see mounting public frustration at prevailing patterns of maltreatment and neglect. And we see growing public fascination with brain science that spills over into overconfidence in psychiatric drugs. These things suggest a strong public interest in finding ways to effectively resolve the current mess, and an underlying public desire for moves directed at promoting better care.

Psychiatry cannot say, as the embattled Winston Churchill said in 1941, "Give us the tools and we will finish the job." Psychiatry's own split personality is not yet healed, and so it is not quite ready to move ahead like that. But the knowledge needed to move ahead is at hand, and patience with the neglect of the mentally ill is wearing thin. So there is compelling reason for psychiatry to adopt a new brain science approach that will be essentially neurodynamic. For it is this and only this that will ultimately enable psychiatry to reawaken public confidence, call up new funds, provide vast social benefits, revolutionize the handling of mental ills, and realize the ancient dream of effectively treating the brain and mind as one.

Notes

Chapter 2

1. National Advisory Mental Health Council, *American Journal of Psychiatry* 150(1993): 1447–1465.
2. E. F. Torrey, *Out of the Shadows*, p. 70.
3. Ibid., p. 40.
4. F. Butterfield, *The New York Times*, July 12, 1999.

Chapter 3

1. E. Shorter, *A History of Psychiatry*, p. 92.

Chapter 4

1. R. J. Isaac and V. C. Armat, *Madness in the Streets*, p. 44.
2. L. V. Briggs et al., *History of the Boston Psychopathic Hospital*, p. 161.
3. E. Shorter, *A History of Psychiatry*, p. 311.

Chapter 5

1. R. W. Sperry, *Science* 113(1961): 1749.

Chapter 8

1. J. LeDoux, *The Emotional Brain*, pp. 168–169.
2. J. E. Brody, *The New York Times*, March 21, 2000, p. D8.
3. National Institute of Mental Health, *The Numbers Count*, p. 2.

Chapter 9

1. J. D. Reed, *People Weekly,* September 13, 1996, p. 44.
2. Idem.

Chapter 10

1. J. Leonard, *Harvard Magazine,* May–June 1999, p. 60.

Chapter 12

1. T. M. Luhrmann, *Of Two Minds,* p. 102.
2. Ibid., pp. 110–111.

Chapter 13

1. E. Roskes, *Psychiatric Services,* 50(1999): 1596.
2. J. B. Gerbasi et al., *Journal of the American Academy of Psychiatry and Law* 28(2000): 128.

Selected Sources

Books

Andreasen, Nancy C. *The Broken Brain: The Biological Revolution in Psychiatry.* New York: Harper and Row, 1984.

Applebaum, Paul S. *Almost a Revolution: Mental Health Law and the Limits of Change.* New York: Oxford University Press, 1994.

Barondes, Samuel H. *Molecules and Mental Illness.* New York: Scientific American Library, 1993.

Briggs, L. Vernon, and collaborators. *History of the Boston Psychopathic Hospital, Boston, Massachusetts.* Boston: Wright and Potter Printing Company, 1922.

Calvin, William H. *The Cerebral Symphony.* New York: Bantam, 1990.

Carter, Rita. *Mapping the Mind.* Berkeley: University of California Press, 1998.

Cohen, Robert I., and Jerome J. Hart. *Student Psychiatry Today: A Comprehensive Textbook.* Oxford: Heinemann Professional Publishing, 1988.

Coren, Stanley, Clare Porac, and Lawrence M. Ward. *Sensation and Perception,* 2nd ed. Orlando, Fla.: Academic Press, 1984.

Crick, Francis. *The Astonishing Hypothesis: The Scientific Search for the Soul.* New York: Simon and Schuster, 1995.

Edelman, Gerald M. *The Remembered Present: A Biological Theory of Consciousness.* New York: Basic Books, 1989.

Glenmullen, Joseph. *Prozac Backlash.* New York: Simon and Schuster, 2000.

Goisman, R. M., and Byrnes, C. *Residency Training on Multidisciplinary Teams at the Massachusetts Mental Health Center.* Proceedings of the American Association of Chairmen of Department of Psychiatry–American Association of Directors of Psychiatry Residency Training Conference on the Impact of Economic and Health

Care Delivery Changes on Psychiatric Residency Training, August 15–17, 1995. Baltimore: University of Maryland Department of Psychiatry, 1995.

Goodwin, Frederick K., and Kay Redfield Jamison. *Manic-Depressive Illness.* New York: Oxford University Press, 1990.

Gutheil, Thomas G., and Paul S. Applebaum. *Clinical Handbook of Psychiatry and the Law,* 3rd ed. Philadelphia: Lippincott Williams and Wilkins, 2000.

Hobson, J. Allan. *Consciousness.* New York: Scientific American Library, 1999.

Hobson, J. Allan. *The Dream Drugstore.* Cambridge, Mass.: MIT Press, 2001.

Hobson, J. Allan. *The Dreaming Brain.* New York: Basic Books, 1988.

Hobson, J. Allan. *Dreaming as Delirium.* Cambridge, Mass.: MIT Press, 2000.

Hobson, J. Allan. *Sleep.* New York: Scientific American Library, 1989 and 1995.

Hunt, Morton. *The Story of Psychology.* New York: Anchor Books/Doubleday, 1993.

Isaac, Rael Jean, and Virginia C. Armat. *Madness in the Streets.* New York: The Free Press/Macmillan, 1990.

James, William. *The Principles of Psychology.* Cambridge, Mass.: Harvard University Press, 1983 edition.

Jones, Edward G. *The Thalamus.* New York: Plenum Press, 1985.

Kramer, Peter D. *Listening to Prozac.* New York: Viking, 1993.

LeDoux, Joseph. *The Emotional Brain.* New York: Simon and Schuster, 1998.

Luhrmann, T. M. *Of Two Minds: The Growing Disorder in American Psychiatry.* New York: Alfred A. Knopf, 2000.

Macphail, Euan M. *The Neuroscience of Animal Intelligence.* New York: Columbia University Press. 1993.

Morris, R. G. M., ed. *Parallel Distributed Processing: Implications for Psychology and Neurobiology.* Oxford: Clarendon Press, 1989.

Nadeau, Robert L. *Mind, Machines, and Human Consciousness.* Chicago: Contemporary Books, 1991.

National Institute of Mental Health. *The Numbers Count: Mental Illness in America* (Publication No. NIH 99–4584). Washington, D.C.: National Institutes of Health, 1999.

Ornstein, Robert. *The Evolution of Consciousness.* New York: Prentice-Hall, 1991.

Ornstein, Robert. *The Psychology of Consciousness.* San Francisco: W. H. Freeman, 1972.

Perrine, Daniel M. *The Chemistry of Mind-Altering Drugs: History, Pharmacology, and Cultural Context.* Washington, D.C.: American Chemical Society, 1996.

Schmahmann, Jeremy D., ed. *The Cerebellum and Cognition.* San Diego: Academic Press, 1997.

Shore, M. F., and Gudeman, J. E. *Serving the Chronically Mentally Ill in an Urban Setting.* New Directions for Mental Health Services, No. 39. San Francisco: Jossey-Bass, 1988.

Shorter, Edward. *A History of Psychiatry: From the Era of the Asylum to the Age of Prozac.* New York: John Wiley & Sons, 1997.

Spitzer, Robert L., and Donald F. Klein, eds. *Evaluation of Psychological Therapies: Psychotherapies, Behavior Therapies, Drug Therapies, and Their Interactions.* Baltimore: The Johns Hopkins University Press, 1976.

Steriade, M., E. G. Jones, and D. A. McCormick. *Thalamus* (2 volumes). Amsterdam: Elsevier, 1997.

Sullivan, Donald M. *The Hospital at 74 Fenwood Road: Boston Psychopathic Hospital.* Boston: Boston Psychopathic Hospital, 1949.

Torrey, E. Fuller. *The Death of Psychiatry.* Radnor, Penn.: Chilton Book Company, 1974.

Torrey, E. Fuller. *Nowhere to Go: The Tragic Odyssey of the Homeless Mentally Ill.* New York: Harper and Row, 1988.

Torrey, E. Fuller. *Out of the Shadows: Confronting America's Mental Illness Crisis.* New York: John Wiley & Sons, 1997.

Tsuang, Ming T., and Stephen V. Faraone. *Schizophrenia, the Facts.* New York: Oxford University Press, 1997.

Articles

Andersson C., M. Chakos, R. Mailman, and J. Lieberman. "Emerging Roles for Novel Antipsychotic Medications in the Treatment of Schizophrenia," *Psychiatric Clinics of North America* 21(1998): 151–179.

Andreasen, Nancy C. "Body and Soul" (editorial), *American Journal of Psychiatry* 153(1996): 589–590.

Andreasen, Nancy C. "The Evolving Concept of Schizophrenia: From Kraepelin to the Present and Future," *Schizophrenia Research* 28(1997): 105–109.

Andreasen, Nancy C. "Linking Mind and Brain in the Study of Mental Illnesses: A Project for a Scientific Psychopathology," *Science* 275(1997): 1586–1593.

Andreasen, Nancy C. "The Role of the Thalamus in Schizophrenia," *Canadian Journal of Psychiatry* 42(1997): 27–33.

Andreasen, Nancy C. "Understanding Schizophrenia: A Silent Spring?" (editorial), *American Journal of Psychiatry* 155(1998): 1657–1659.

Andreasen, N. C., D. S. O'Leary, T. Cizadlo, S. Arndt, K. Rezai, L. L. Ponto, G. L. Watkins, and R. D. Hichwa. "Schizophrenia and Cognitive Dysmetria: A Positron-Emission Tomography Study of Dysfunctional Prefrontal-Thalamic-Cerebellar Circuitry," *Proceedings of the National Academy of Sciences* 93(1996): 9985–9990.

Andreasen, Nancy C., Sergio Paradiso, and Daniel S. O'Leary. "'Cognitive Dysmetria' as an Integrative Theory of Schizophrenia: A Dysfunction in Cortical-Subcortical-Cerebellar Circuitry?" *Schizophrenia Bulletin* 24(1998): 203–218.

Applebaum, Paul S. "Managed Care and the Next Generation of Mental Health Law," *Psychiatric Services* 47(1996): 27.

Barloon, Thomas J., and Russell Noyes, Jr. "Charles Darwin and Panic Disorder," *Journal of the American Medical Association* 277(1997): 138–141.

Barnes, T. R., and M. A. McPhillips, "Novel Antipsychotics, Extrapyramidal Side Effects, and Tardive Dyskinesia," *International Clinical Psychopharmacology* 13, suppl 3(1998): S49–S57.

Blakeslee, Sandra. "For Better Learning, Researchers Endorse 'Sleep on It' Adage," *New York Times*, March 7, 2000.

Cloud, John. "The Lure of Ecstasy," *Time*, June 5, 2000.

Cloud, John. "Mental Health Reform: What It Would Really Take," *Time*, June 7, 1999.

Cohen, Bruce M., and Weihua Wan. "The Thalamus as a Site of Action of Antipsychotic Drugs" (brief reports), *American Journal of Psychiatry* 153(1995): 104–106.

Crespo-Facorro, B., Sergio Paradiso, N. C. Andreasen, D. S. O'Leary, G. L. Watkins, L. L. Boles Ponto, and R. D. Hichwa. "Recalling Word Lists Reveals 'Cognitive Dysmetria' in Schizophrenia: A Positron Emission Tomography Study," *American Journal of Psychiatry* 156(1999): 386–392.

Ditton, Paula M. *Mental Health and Treatment of Inmates and Probationers* (special report). Washington, D.C.: U.S. Bureau of Justice Statistics, 1999.

Drevets, W. C., D. Ongur, and J. L. Price. "Neuroimaging Abnormalities in the Subgenual Prefrontal Cortex: Implications for the Pathophysiology of Familial Mood Disorders," *Molecular Psychiatry* 3(1998): 190–191.

Drevets, W. C., J. L. Price, J. R. Simpson, Jr., R. D. Todd, T. Reich, M. Vannier, and M. E. Raichle. "Subgenual Prefrontal Cortex Abnormalities in Mood Disorders," *Nature* 386(1997): 769–770.

Druss, Benjamin G., and Robert A. Rosenheck. "Mental Disorders and Access to Medical Care in the United States," *American Journal of Psychiatry* 155(1998): 1775–1777.

Duncan, G. E., B. B. Sheitman, and J. A. Lieberman. "An Integrated View of Pathophysiological Models of Schizophrenia," *Brain Research Reviews* 29(1999): 250–264.

Egan, Michael F., and Daniel R. Weinberger. "Neurobiology of Schizophrenia," *Current Opinion in Neurobiology* 7(1997): 701–707.

Eichenbaum, Howard. "How Does the Brain Organize Memories?" *Science* 277(1997): 330–331.

Eist, Harold I. "Treatment for Major Depression in Managed Care and Fee-for-Service Systems" (editorial), *American Journal of Psychiatry* 155(1998): 859–860.

Ernst, E., and A. Herxheimer. "The Power of Placebo: Let's Use It to Help as Much as Possible," *British Medical Journal* 313(1996): 1569.

Gabbard, Glen O. "The Impact of Psychotherapy on the Brain," *Psychiatric Times,* September 1998.

Gerbasi, Joan B., Richard J. Bonnie, and Renée L. Binder. "Resource Document on Mandatory Outpatient Treatment," *Journal of the American Academy of Psychiatry and Law* 28(2000): 127–143.

Glausiusz, Josie. "The Chemistry of Obsession: Behavior Therapy for Obsessive-Compulsive Disorder Changes Brain Chemistry," *Discover,* June 1996.

Goldman-Rakic, Patricia S. "Working Memory Dysfunction in Schizophrenia," >*Journal of Neuropsychiatry and Clinical Neuroscience* 6(1994): 348–357.

Gorman, Christine. "Anatomy of Melancholy: Scientists Take a Picture of Depression and Discover That It Actually Changes the Shape of the Brain," *Time,* May 5, 1997.

Green, Alan I., and Joseph J. Schildkraut. "Should Clozapine Be a First-line Treatment for Schizophrenia? The Rationale for a Double-blind Clinical Trial in First-Episode Patients," *Harvard Review of Psychology* 1 (1995): 1–9.

Grinfeld, Michael Jonathan. "From Poster Child to Wanted Poster: Explaining the Realities of Mental Illness," *Psychiatric Times,* September 1998.

Hall, Stephen S. "Fear Itself: What We Now Know About How It Works, How It Can Be Treated, and What It Tells Us about Our Unconscious," *New York Times Magazine,* February 28, 1999.

Hirschfeld, Robert M. A., and James M. Russell. "Assessment and Treatment of Suicidal Patients," *New England Journal of Medicine* 337(1997): 910–915.

Hobson, J. Allan. "How the Brain Goes Out of Its Mind," *Endeavour,* June 1996.

Hobson, J. Allan. "The New Neuropsychology of Sleep: Implications for Psychoanalysis," *Neuropsychoanalysis* 1(1999): 157–183.

Hobson, J. Allan, and Rosalia Silvestri. "Sleep Disorders." In Armand Nicholi, ed. *Harvard Guide to Psychiatry,* 2nd ed. Cambridge, Mass.: Harvard University Press, 1999.

Hobson, J. Allan, Robert Stickgold, and Edward F. Pace-Schott. "The Neuropsychology of REM Sleep Dreaming," *NeuroReport* 9(1998): R1-R14.

Hughes, Douglas H. "Can the Clinician Predict Suicide?" *Psychiatric Services* 46(1995): 449–451.

Husted, June R. "Insight in Severe Mental Illness: Implications for Treatment Decisions," *Journal of the American Academy of Psychiatry and Law* 27(1999): 33–49.

Kandel, Eric R. "Biology and the Future of Psychoanalysis: A New Intellectual Framework for Psychiatry Revisited," *American Journal of Psychiatry* 156(1999): 505–524.

Kandel, Eric R. "A New Intellectual Framework for Psychiatry," *American Journal of Psychiatry* 155(1998): 457–469.

Kandel, Minouche, and Eric Kandel. "Flights of Memory" (the biology of recovered memory, column), *Discover*, May 1994.

Keefe, Richard S. E., Susan G. Silva, Diana O. Perkins, and J. A. Lieberman. "The Effects of Atypical Antipsychotic Drugs on Neurocognitive Impairment in Schizophrenia: A Review and Meta-analysis," *Schizophrenia Bulletin* 25(1999): 201–222.

Kendler, Kenneth S. "Long-Term Care of an Individual with Schizophrenia: Pharmacologic, Psychological, and Social Factors," *American Journal of Psychiatry* 156(1999): 124–128.

Kramer, Mark S., N. Cutler, J. Feighner, R. Shrivastava, J. Carman, J. J. Sramek, S. A. Reines, G. Liu, D. Snavely, E. Wyatt-Knowles, J. J. Hale, S. G. Mills, M. MacCoss, C. J. Swain, T. Harrison, R. G. Hill, F. Hefti, E. M. Scolnick, M. A. Cascieri, G. G. Chicchi, S. Sadowski, A. R. Williams, L. Hewson, D. Smith, Nadia M. J. Rupniak, *et al.* "Distinct Mechanism for Antidepressant Activity by Blockade of Central Substance P Receptors," *Science* 281(1998): 1640–1645.

Leonard, Jonathan. "Dream-Catchers: Unleashing the Genies in the Sleeping Mind," *Harvard Magazine,* May-June 1998.

Leonard, Jonathan. "The Sorcerer's Apprentice: Unlocking the Secrets of the Brain's Basement," *Harvard Magazine,* May-June 1999.

Leonard, Jonathan. "William Gorgas, Soldier of Public Health," *Bulletin of the Pan American Health Organization* 25(1991): 166–185.

Lieberman, Jeffrey A. "Maximizing Clozapine Therapy: Managing Side Effects," *Journal of Clinical Psychiatry* 59, suppl 3(1998): 38–43.

Lieberman, Paul B., Stephen A Wiitala, Binette Elliott, Sandra McCormick, and Stephanie B. Goyette. "Decreasing Length of Stay: Are There Effects on Outcomes of Psychiatric Hospitalization?" *American Journal of Psychiatry* 155(1998): 905–909.

Llinás, Rodolfo R., Urs Ribary, Daniel Jeanmonod, Eugene Kronberg, and Partha B. Mitra. "Thalamocortical Dysrhythmia: A Neurological and Neuropsychiatric Syndrome Characterized by Magnetoencephalography," *Proceedings of the National Academy of Sciences* 96(1999): 15222–15227.

Marzuk, Peter M., and Jack D. Barchas. "Psychiatry," *Journal of the American Medical Association* 277(1997): 1892.

Marzuk, Peter M., Kenneth Tardiff, Andrew C. Leon, Charles S. Hirsch, Marina Stajic, Nancy Hartwell, and Laura Portera. "Use of Prescription Psychotropic Drugs among Suicide Victims in New York City," *American Journal of Psychiatry* 152(1995): 1520–1522.

McFadden, Robert D. "New York Nightmare Kills a Dreamer," *New York Times,* January 4, 1999.

Miller, C. H., F. Mohr, D. Umbricht, M. Woerner, W. W. Fleischhacker, and J. A Lieberman. "The Prevalence of Acute Extrapyramidal Signs and Symptoms in Patients Treated with Clozapine, Risperidone, and Conventional Antipsychotics," *Journal of Clinical Psychiatry* 59(1998): 69–75.

Nash, J. Madeleine. "Fertile Minds," *Time,* February 3, 1997.

National Advisory Mental Health Council. "Health Care Reform for Americans with Severe Mental Illnesses: Report of the National Advisory Mental Health Council," *American Journal of Psychiatry* 150(1993): 1447–1465.

Nopoulos, P. C., J. W. Ceilley, E. A. Gailis, and N. C. Andreasen. "An MRI Study of Cerebellar Vermis Morphology in Patients with Schizophrenia: Evidence in Support of the Cognitive Dysmetria Concept," *Biological Psychiatry* 46(1999): 703–711.

Nutt, David. "Substance-P Antagonists: A New Treatment for Depression?" *Lancet* 352(1998): 1644.

Nutt, David J. "Addiction: Brain Mechanisms and Their Treatment Implications," *Lancet* 347(1996): 31–36.

Olfson, Mark, Steven C. Marcus, and Harold Alan Pincus. "Trends in Office-Based Psychiatric Practice," *American Journal of Psychiatry* 156(1999): 451–457.

Olfson, M., D. Mechanic, S. Hansell, C. A. Boyer, and J. Walkup. "Predictions of Homelessness within Three Months of Discharge among Inpatients with Schizophrenia," *Psychiatric Services* 50(1999): 667–673.

Olfson, Mark, Terri L. Tanielian, Brennan D. Peterson, and Deborah A. Zarin. "Routine Treatment of Adult Depression," *Psychiatric Services* 49(1998): 299.

Ongur, D., W. C. Drevets, and J. L. Price. "Glial Reduction in the Subgenual Prefrontal Cortex in Mood Disorders," *Proceedings of the National Academy of Sciences* 95(1998): 13290–13295.

Pace-Schott, Edward F., Tamara Gersh, Rosalia Silvestri, Robert Stickgold, Carl Salzman, and J. Allan Hobson. "SSRI Treatment Suppresses Dream Recall Frequency but Increases Subjective Dream Intensity in Normal Subjects," submitted to *Journal of Sleep Research.*

Pardes, Herbert. "Future Needs for Psychiatrists and Other Mental Health Personnel," *Archives of General Psychiatry* 36(1979): 1401–1408.

Patel, Jayendra K, Carl Salzman, Alan I. Green, and Ming T. Tsuang. "Chronic Schizophrenia: Response to Clozapine, Risperidone, and Paroxetine," *American Journal of Psychiatry* 154(1997): 543–546.

Penn, David L., and Kim T. Mueser. "Research Update on the Psychosocial Treatment of Schizophrenia," *American Journal of Psychiatry* 153(1996): 607–617.

Perry, W., and D. L. Braff. "Information-Processing Deficits and Thought Disorder in Schizophrenia," *American Journal of Psychiatry* 151(1994): 363–367.

Petrila, John. "Who Will Pay for Involuntary Civil Commitment under Capitated Managed Care? An Emerging Dilemma," *Psychiatric Services* 46(1995): 1045–1048.

Powledge, Tabitha M. "Unlocking the Secrets of the Brain," *BioScience* 47(1997): 403–409.

Raedler, T. J., M. B. Knable, and D. R. Weinberger. "Schizophrenia as a Developmental Disorder of the Cerebral Cortex," *Current Opinion in Neurobiology* 8(1998): 157–161.

Rafal, Seth, Ming T. Tsuang, and William T. Carpenter, Jr. "A Dilemma Born of Progress: Switching from Clozapine to a Newer Antipsychotic," *American Journal of Psychiatry* 156(1999): 1086–1090.

Rapoport, Judith L., and Alan Fiske. "The New Biology of Obsessive-Compulsive Disorder: Implications for Evolutionary Psychology," *Perspectives in Biology and Medicine* 41(1998): 159–175.

Robbins, Jim. "Wired for Sadness," *Discover,* April 2000.

Robinson, Delbert, Margaret G. Woerner, Jose Ma. J. Alvir, Robert Bilder, Robert Goldman, Stephen Geisler, Amy Koreen, Brian Sheitman, Miranda Chakos, David Mayerhoff, and Jeffrey A. Lieberman. "Predictors of Relapse following Response from a First Episode of Schizophrenia or Schizoaffective Disorder," *Archives of General Psychiatry* 56(1999): 241–247.

Roskes, Erik. "Offenders with Mental Disorders: A Call to Action," *Psychiatric Services* 50(1999): 1596.

Salzman, Carl. "Integrating Pharmacotherapy and Psychotherapy in the Treatment of a Bipolar Patient," *American Journal of Psychiatry* 155(1998): 686–688.

Schildkraut, Joseph J., and Seymour S. Kety. "Biogenic Amines and Emotion," *Science* 156 (1967): 21–37

Schlozman, Steve. "The Bard in His Outcast State," *Journal of the American Medical Association* 277(1997): 1118.

Schmahmann, Jeremy D. "Dysmetria of Thought: Clinical Consequences of Cerebellar Dysfunction on Cognition and Affect," *Trends in Cognitive Sciences* 2(1998): 362–371.

Schmahmann, Jeremy D. "An Emerging Concept: the Cerebellar Contribution to Higher Thought," *Archives of Neurology* 48(1991): 1178–1187.

Schultz, Susan K., and Nancy C. Andreasen. "Schizophrenia," *Lancet* 353(1999): 1425–1430.

Sheehan, Susan. "A Reporter at Large." *The New Yorker Magazine,* May 25, June 1, June 8, and June 15, 1981.

Shorter, Edward. "Prozac vs. Freud: Medicine Wins," *Current,* January 1999.

Sierles, Frederick S., and Michael Alan Taylor. "Decline of U.S. Medical Student Career Choice of Psychiatry and What to Do About It," *American Journal of Psychiatry* 152(1995): 1416–1426.

Silvestri-Hobson, Rosalia, Edward F. Pace-Schott, Tamara Gersh, Robert Stickgold, Carl Salzman, and J. Allan Hobson. "Effects of Fluvoxamine and Paroxetine on Sleep Structure in Normal Subjects: A Home Based Nightcap Evaluation during Drug Administration and Withdrawal," *Journal of Clinical Psychology* (in press).

Sperry, R. W. "Cerebral Organization and Behaviour," *Science* 133(1961): 1749–1757.

Suominen, Kirsi H., Erkki T. Isometsä, Markus M. Henriksson, Aini I. Ostamo, and Jouko L. Lönnqvist. "Inadequate Treatment for Major Depression both before and after Attempted Suicide," *American Journal of Psychiatry* 155(1998): 1778–1780.

Swanson, Jeffrey W., Marvin S. Swartz, Randy Borum, Virginia A. Hiday, H. Ryan Wagner, and Barbara J. Burns. "Involuntary Out-patient Commitment and Reduction of Violent Behaviour in Persons with Severe Mental Illness," *British Journal of Psychiatry* 176(2000): 324–331.

Terry, Ken. "Where's Managed Care Headed?" *Medical Economics*, April 10, 2000.

Tononi, Giulio, and Gerald M. Edelman. "Consciousness and Complexity," *Science* 282(1998): 1846–1851.

Torrey, E. Fuller. "Why Are There So Many Homeless Mentally Ill?" *Harvard Medical School Mental Health Letter*, August 1989.

Torrey, E. Fuller, and Robert J. Kaplan. "A National Survey on the Use of Outpatient Commitment," *Psychiatric Services* 46(1995): 778–784.

Tran, Khoa D., Gregory S. Smutzer, Richard L. Doty, and Steven E. Arnold. "Reduced Purkinje Cell Size in the Cerebellar Vermis of Elderly Patients with Schizophrenia," *American Journal of Psychiatry* 155(1998):1288–1290.

Tsuang, Ming T. "Genetic Epidemiology of Schizophrenia: Review and Reassessment," *Kao Hsiung I Hsueh Ko Hsueh Tsa Chih* 14(1998): 405–412.

Tsuang, Ming T. "The Massachusetts Mental Health Center" (images in psychiatry), *American Journal of Psychiatry* 154(1997): 423.

Tsuang, Ming T. "Schizophrenia: Genes and Environment," *Biological Psychiatry* 47(2000): 210–220.

Tsuang, Ming T., and Stephen V. Faraone. "Genetic Heterogeneity of Schizophrenia," *Seishin Shinkeigaku Zasshi* 97(1995): 485–501.

Tsuang, Ming T., William S. Stone, Larry J. Seidman, Stephen V. Faraone, Suzanna Zimmet, Joanne Wojcik, James P. Kelleher, and Alan I. Green. "Treatment of Nonpsychotic Relatives of Patients with Schizophrenia: Four Case Studies," *Biological Psychiatry* 45(1999): 1412–1418.

Tucker, W. "The 'Mad' vs. the 'Bad' Revisited: Managing Predatory Behavior," *Psychiatric Quarterly* 70(1999): 221–230.

Wassink, T. H., N. C. Andreasen, P. Nopoulos, and M. Flaum. "Cerebellar Morphology as a Predictor of Symptom and Psychosocial Outcome in Schizophrenia," *Biological Psychiatry* 45(1999): 41–48.

Weinberger, D. R., K. F. Berman, R. Suddath, and E. F. Torrey. "Evidence of Dysfunction of a Prefrontal–Limbic Network in Schizophrenia: A Magnetic Resonance Imaging and Regional Cerebral Blood Flow Study of Discordant Monozygotic Twins," *American Journal of Psychiatry* 149(1992): 890–897.

Weinberger, D. R., and B. Gallhofer. "Cognitive Function in Schizophrenia," *International Clinical Psychopharmacology* 12, suppl 4(1997): S29-S36.

Weinberger, Daniel R. "The Biological Basis of Schizophrenia: New Directions," *Journal of Clinical Psychiatry* 58, suppl 10(1997): 22–27.

Wickelgren, Ingrid. "Getting a Grasp on Working Memory," *Science* 275(1997): 1580–1582.

Willwerth, James. "Working Their Way Back," *Time*, November 22, 1999.

Yager, Joel. "Psychiatric Residency Training and the Changing Economic Scene," *Hospital and Community Psychiatry* 38(1987): 1076–1081.

Index